Indian Herbal Drug Microscopy

Shailendra S. Gurav • Nilambari S. Gurav
Editors

Indian Herbal Drug Microscopy

Springer

Editors
Shailendra S. Gurav
Department of Pharmacognosy
Government College of Pharmacy
Karad, Maharashtra, India

Nilambari S. Gurav
Department of Pharmacognosy
Sudhakarrao Naik Institute of Pharmacy
Pusad, Maharashtra, India

ISBN 978-1-4614-9514-7 ISBN 978-1-4614-9515-4 (eBook)
DOI 10.1007/978-1-4614-9515-4
Springer New York Heidelberg Dordrecht London

Library of Congress Control Number: 2013954404

Printed on acid-free paper

Springer is part of Springer Science+Business Media (www.springer.com)

Dedicated to our beloved Aai, Anna, and Anu

Foreword

The use of plants as a source of medicine is as old as humanity. Medicinal plants have always played a significant role in treating illness or preventing disease. Over the past several decades, scientific literature and popular media articles on adverse drug effects increased the interest in natural products by the general public. Major issues in plant drug research are proper identification, authentication, and evaluation of the plants. Pharmacognostical and phytochemical evaluation plays a major role in this context.

Anatomical study of plants plays a vital role for their identification and authentication. The knowledge of microscopic details of plants in crude and powder form is vital for the evaluation of medicinal plants in every way. This book on "Indian Herbal Drug Microscopy" will be very helpful in this context for the identification and evaluation of medicinal plants. The book contains several chapters ranging from stepwise procedure for sectioning of plant material to histochemical staining techniques and anatomy of 40 well-known and medicinally important plants like Arjuna, Ashoka, Ashwagandha, Cinchona, Cinnamon, Ginger, Kurchi, Rauwolfia, Turmeric, Tulsi, and Vasaka with hand-drawn colored microscopic images of crude drugs and their magnified powder microscopic characters.

I appreciate the efforts put forward by the authors to develop this laboratory manual, which is an outcome of their experience and hard work. I hope this not only will be very helpful for students but also will be an essential reference for anyone involved in the fields of phytomedicine, traditional herbal remedies, pharmaceutical sciences, and natural product research.

Kolkata, India — Pulok K. Mukherjee, Ph.D., F.R.S.C.

Preface

The past decade has witnessed the introduction and implementation of current Good Manufacturing Practices (GMP) in quality control of raw materials, intermediates, and finished products of botanical origin to overcome adulteration and misidentification of herbal drugs. The initial step in quality control of medicinal plants is confirming the authenticity of the desired species via a variety of techniques such as macro- and microscopic identification and chemical analysis. The botanical control by microscopic examinations, using histological identification, is still most preferred as a rapid and inexpensive technique.

Today there are few laboratory manuals and practical handbooks which highlight microscopy of crude drugs only. However, our book differentiates from others in the context of hand-drawn colored microscopic images of crude drugs and their powder characters as observed under a microscope after magnification. There may be many omissions and commissions, so the author sincerely wishes the kind cooperation of every reader in correcting them.

This work is the output of a number of years of experience in training imparted to pharmacy students in the identification of herbal drugs. It describes today's knowledge of some important botanical microscopic characters of the whole, fragmented, and powdered herbal drugs studied. While referring to this book to authenticate a given sample, it will be necessary to make microscopic preparations of the plant material under study in order to compare the structures present with those drawn and described herewith. All the drawings in this book have been made by us after observation of microscopic preparations by standard techniques from previously authenticated samples, in our laboratory. In preparing the drawings our purpose is to illustrate all the diagnostic as well as microscopic characters which play a crucial role in authentication of plant species. The color diagrams are sufficiently clear, exemplifying that anyone can easily match the characters seen under the microscope with the drawn diagrams. Practical aspects of sectioning and histochemical staining techniques are also described as separate chapters. It also makes the book user friendly for analysts working in pharmaceutical concerns manufacturing herbal products.

I hope our present attempt will be helpful for students, teachers, and researchers of pharmacy and *Ayurveda* as well as analysts in the herbal and Ayurvedic industry.

Karad, Maharashtra, India Shailendra Gurav
Pusad, Maharashtra, India Nilambari Gurav

Acknowledgements

It is always hard to list comprehensively all those whose influence has been felt and whose assistance has been liberally sought in the gestation of this manuscript. This book is an output of a number of years of experience in training imparted to pharmacy students in the identification of herbal drugs.

The editors are particularly grateful to Dr. Arun Patil, Dr. Shrikant Tilloo, and Dr. Kishorkumar Burade for their association with the work at various stages and for the helpful comments during the process. The editors warmly appreciate the arduous and dedicated artwork put forth by Mr. Atul Ahire without whom this book was never possible. The editors are also thankful to Dr. Asha Arora, Head of Biotechnology Department, B.N.P.G. College, Udaipur, Rajasthan, India; Mrs. Shubhada Nikharge, Founder Trustee of "Save Rani Bagh Botanical Garden Foundation," Mumbai, India; and Dr. Sandip Hate for their help by providing actual photos of plants for this manuscript. We are very much thankful to Dr. Pulok Mukherjee, Director, School of Natural Product Studies, Jadavpur University, Kolkata, for his appreciation and recommendation in the form of "Foreword" for our book. The editors take this opportunity to thank Dr. Dadasaheb Kokare, Dr. Prashant Puranik, Dr. Vijay Gulkari, and Dr. Santosh Bhujbal for their support at every stage in the preparation of this manuscript. The editors like to express their gratitude to all their colleagues working in different departments for their timely help and moral support.

Finally, the editors extend their appreciation to their parents Aai, Anna, and Baba without whose blessings it was never possible to bring this dream into reality and their little princess "Anu" to whom this book is dedicated.

Karad, Maharashtra, India — Shailendra Gurav
Pusad, Maharashtra, India — Nilambari Gurav

Content

1 **Introduction** 1
Shailendra Gurav and Nilambari Gurav

2 **Sectioning Methods** 5
Shailendra Gurav, Nilambari Gurav, and Arun Patil

3 **Histological and Histochemical Staining Techniques** 9
Shailendra Gurav, Shrikant Tilloo, and Kishor Burade

4 **Herbal Drug Microscopy** 15
Shailendra Gurav and Nilambari Gurav

Bibliography 197

Index 199

Contributors

Kishor Burade Government College of Pharmacy, Karad, Maharashtra, India

Nilambari S. Gurav Sudhakarrao Naik Institute of Pharmacy, Pusad, Maharashtra, India

Shailendra S. Gurav Government College of Pharmacy, Karad, Maharashtra, India

Arun Patil Department of Pharmaceutical Sciences, Rashtrasant Tukadoji Maharaj University, Nagpur, Maharashtra, India

Shrikant Tilloo Gurunanak College of Pharmacy, Nagpur, Maharashtra, India

Chapter 1
Introduction

Shailendra Gurav and Nilambari Gurav

The use of herbal medicines throughout the world has become more and more popular in recent years. Among these, India and China are the most upcoming and blooming countries right now. Extensive use of herbal medicine poses two new problems, with international implications, which increase the importance of fast, accurate, and efficient means of authenticating herbs. First, the growing market for herbal medicines worldwide has engendered many international trading companies and generated an increase in counterfeit herbs and herbs of questionable quality. Second, herbal medicines are often taken as combinations (most of the time) which generate unique problems of authentication, such as determining if there is species confusion of different herbs sharing one name or one herb using different names and if the correct herbal medicine has been included in a particular proprietary medicine.

Adulteration and misidentification of herbal drug can cause serious health problems to consumers, as well as publicity and legal headaches for the pharmaceutical industry. Many poisoning incidents caused by misuse or confusion of herbal medicines have raised international concern for authentication of herbal medicines to their safe and effective use. The past decade has witnessed the introduction and institution of current Good Manufacturing Practices (GMP) in quality control of raw materials, intermediates, and finished products of botanical origin.

The first step in quality control of medicinal plants is ensuring the authenticity of the herbal medicines, and today, there are a variety of methods available to authenticate herbal medicines, ranging from simple morphological examination to physical and chemical analysis and DNA molecular biology. However, each method

S. Gurav (✉)
Government College of Pharmacy, Karad, Maharashtra, India
e-mail: shailugurav@rediffmail.com

N. Gurav
Sudhakarrao Naik Institute of Pharmacy, Pusad, Maharashtra, India
e-mail: nilagurav@rediffmail.com

S.S. Gurav and N.S. Gurav (eds.), *Indian Herbal Drug Microscopy*,
DOI 10.1007/978-1-4614-9515-4_1, © Springer Science+Business Media New York 2014

has drawbacks and advantages. In difficult or critical cases, sometimes two or several methods are applied for the authentication of herbal medicines.

Microscopy permits the identification of herbal drugs and the detection of individual components of a mixture by examining their unique features like histological structures, cells, and cell contents. It is of great value in case of comparative analysis of broken and powdered herbal products because in such cases most of the morphological diagnostic features are lost. The powdered crude drugs can be identified based on the form, the presence, or the absence of different cell types based on their cytomorphological characters, e.g., parenchyma, collenchyma, fibres, stone cells, vessels, trichomes, secretory cells, and epidermal cells.

For several years, the magnifying glass and the microscope were the only possible methods for the analytical evaluation of herbal drugs. Advances in microscope technology have improved the accuracy and capabilities of microscopy as a tool of botanical identification.

Different types of microscopic techniques which can be used for the pharmacognostic studies include light microscopy (LM), polarizing microscopy, phase-contrast microscopy, and scanning electron microscopy (SEM). However, ordinary light microscopy is still the most common method for primary authentication. It has been commonly used in the authentication of herbal medicines in India and many other countries because of its virtues of requirement of small amount of sample, fast speed, and economy. Furthermore, herbal medicines are mostly low-cost medicine, which should not be raised in price just because of the application of unnecessary highly sophisticated methods for authentication. Indian Ayurvedic Pharmacopoeia (First Edition 1964) and Ayurvedic Formulary of India (First Edition 1966) clearly show that in India microscopic techniques have been used in the authentication of herbal medicines for a long time. Outside India, the pharmacopoeias of other countries also record the microscopic characteristics of their herbal medicines, for example, European Pharmacopoeia, British Pharmacopoeia, United States Pharmacopeia, Japanese Pharmacopoeia, Chinese Pharmacopoeia, and Vietnamese Pharmacopoeia.

Though there are few laboratory manuals and practical handbooks in market which highlight microscopy characters of crude drugs, present book differentiates from others in the form of hand-drawn colour microscopic images of crude drug and its powdered characters as they observed under microscope after magnification. There may be many omissions and commissions, so the author sincerely wishes the kind cooperation of every reader in correcting them.

In the present book, the drugs have been selected taking into consideration the pharmacognosy course of different institutions of pharmacy in India. The drugs have been initially authenticated from the botany department and then microscopy of same was performed in our laboratory by standard techniques. Powdered drugs before use were sieved through a no. 60 mesh and further used for microscopy study.

It describes today's knowledge of some important botanical microscopic characters of the whole, fragmented, and powdered herbal drugs studied. When using the present book to authenticate a given sample, it will be necessary to make microscopic preparations of the material in order to compare the structures present with those drawn and described herewith. All the drawings in the present book have been made by us after observation of prepared microscopic preparations by standard techniques from previously authenticated samples, in our laboratory.

Chapter 2
Sectioning Methods

Shailendra Gurav, Nilambari Gurav, and Arun Patil

1 Introduction

Plant anatomy is a basic core subject in the study of biology, especially plant biology. In the study of plant structure, it is important to recognise that there is a fundamental difference between plant and animal development. Thorough knowledge of the structure of plant cells and tissues is the prerequisite for a realistic interpretation of morphology, physiology, and phylogeny. Moreover, it is also essential to solve many important everyday problems such as the identification of unknowns and food contaminants. Therefore in order to learn about plant structures, it is important to take hands on some of the simple techniques that are useful in the study of plant structures.

2 Necessary Material

The variety of instruments can be used depending upon the nature of work. The microscope is the most indispensable instrument used in laboratory which helps to increase the resolving power of human eye which fails to recognise objects lying closer between 0.01 and 0.25 mm. Dissecting and compound microscopes are most

S. Gurav (✉)
Government College of Pharmacy, Karad, Maharashtra, India
e-mail: shailugurav@rediffmail.com

N. Gurav
Sudhakarrao Naik Institute of Pharmacy, Pusad, Maharashtra, India
e-mail: nilagurav@rediffmail.com

A. Patil
Department of Pharmaceutical Sciences, Rashtrasant Tukadoji Maharaj University, Nagpur, Maharashtra, India
e-mail: profarunpatil@gmail.com

S.S. Gurav and N.S. Gurav (eds.), *Indian Herbal Drug Microscopy*,
DOI 10.1007/978-1-4614-9515-4_2, © Springer Science+Business Media New York 2014

commonly used, whereas binocular, phase-contrast, electron, and fluorescent are other types of microscopes used as per requirement. However, it is more convenient to prepare a small kit containing good and sharp razor blade, a fine hairbrush, watch glasses, Petri dishes, clean slides, coverslips, two fine long handle dissecting needles, a pair of forceps, glass droppers, a pair of sharpened pencils, pencil eraser, a clean and soft handkerchief, etc. Other supplies such as filter papers, lens paper, and lens cleaner for slides and microscope lens should be available in the laboratory.

A set of different staining solutions in dropper bottles, pipettes with rubber bulbs, a large water bottle, and a tray of necessary materials once arranged can be used throughout the technique. The reagents and stains can be replenished as and when required.

3 Freehand Sectioning Methods

Most plant parts are too thick to be mounted intact and viewed with a microscope. In order to study the plant anatomy, sections have to be made so that enough light can be transmitted through the specimen to resolve cell structures under the microscope. A freehand section is the most common and simple method of preparing specimens for microscopic viewing. This method allows one to examine the specimen in a few minutes. It is also suitable for a variety of plant materials, such as soft herbaceous stems and small woody twigs.

Different sections can be obtained from a stem, root, or stolon, depending on the plane of cutting, each section revealing the details from a different angle. Transverse section is obtained by cutting along the radial plane of a cylindrical portion of the stem or root or stolon and perpendicular to the long axis. Tangential longitudinal section is a section cut along the long axis parallel to a tangent, whereas radial longitudinal section is a section cut along the long axis and the cutting plane passing through the long axis and radius.

In general it is found that transverse sections are easier than longitudinal sections as the internal structures are clearly identified in the same. The good transverse sections can be obtained by cutting plant material, to be sectioned at right angle to the long axis of the cells. The fixation of materials is generally not required for temporary preparations. More and more practice, patience, and perhaps one's inherent skill are the prerequisite for this technique.

3.1 Steps

1. Sit comfortably with your forearms resting on the bench and your elbows close to your sides.
2. Have a new sharp double-edged razor blade. To minimise the risk of cutting oneself, cover one edge of the razor blade with masking tape.

3. Hold the plant material firmly against the side of the first finger of any hand (left or right) by means of the thumb. Keep the first finger as straight as possible, while the thumb should kept well below the surface of the material out of the way of the razor edge.
4. Put the drop of water on the razor so as to reduce the friction during cutting as sections can float onto the surface of the blade. Take the razor blade in any hand (left or right) and place it on the first finger of the left hand (or right hand), more or less at a right angle to the specimen.
5. Now move the razor quickly across the top of the material in such manner to give the material a drawing cut in a single complete stroke. This minimises the friction while passing the razor blade through the specimen. Use more and more uniform strokes to get several sections at a time. Sections will certainly vary in thickness. However, there will be usable ones among the thick sections.

   (Note: During sectioning, a number of sections should be cut at the same time without worrying about section thickness at this time. Thick sections are also usable unless they are not obliquely sectioned. One can cut a number of sections without moving the material or the thumb just by slightly and progressively increasing the pressure with the razor blade on the first finger and simultaneously exerting increasing pressure onto the specimen by the thumb. It is advised to start cutting with the razor blade right at the surface of the specimen rather than against the side of the material. In case of the root and stem (as they have a radial symmetry), it is not necessary that a section should be complete, as long as it includes a portion of the tissues from the centre to the outer edge of the specimen.)

6. Transfer sections to water in watch glass using a brush (not a forceps or needle).
7. Select and transfer the thinnest sections (the more transparent ones) onto a clean glass slide, and put two to three drops of chloral hydrate solution on it, and heat the slide very gently by passing to and fro over a low flame. When bubbles start to appear, stop heating, and add a drop of glycerine–water solution to avoid drying of the preparation and crystallisation of chloral hydrate.
8. To apply the coverslip, hold it at an angle and touch the glycerine-water drop with one edge. Lower the coverslip slowly to avoid air bubbles.

   (Note: In case of thin leaves and tiny roots, i.e., delicate and hard to hold specimens, additional support can be used to facilitate hand sectioning. In such cases tissue pieces can be inserted into a small piece of pith such as a potato tuber or carrot or radish root. Once it is sure, the tissue is firmly in place, and the hand sectioning technique can be applied. Without supporting material longitudinal sections are also difficult to obtain as small stem and root pieces are not easy to hold firmly with one's finger. However, by cutting a v-shaped notch into the pith support, sectioning can be done.)

Chapter 3
Histological and Histochemical Staining Techniques

Shailendra Gurav, Shrikant Tilloo, and Kishor Burade

1 Introduction

Section staining is the most captivating part in the preparation of specimens for microscopy. The selected sections need to be stained as stains help to distinguish different tissues, cells or inclusions from one another by developing specific colours.

In general, most biological tissues have very little contrast and it is hard to distinguish cellular details with the ordinary light microscope. Though stains can enhance and improve the visibility of the specimen, different stains have different affinities for various organelles and macromolecules. Therefore, the careful selection and utilisation of stains can also propose the chemical nature of the substances within the cell. Toluidine Blue O (TBO) is most common and an excellent stain for freehand section technique. Phloroglucinol–HCl, iodine–potassium iodide solution and Sudan III or IV are few examples of other stains used for freehand sections according to requirement.

2 Staining Method

1. Keep selected thin sections in a watch glass. Add a few drops of stain to immerse the section.
2. Allow the material to remain for a few minutes so that it will stain (the time required varies with the materials and stain).

S. Gurav (✉) • K. Burade
Government College of Pharmacy, Karad, Maharashtra, India
e-mail: shailugurav@rediffmail.com; k_burade@rediffmail.com

S. Tilloo
Gurunanak College of Pharmacy, Nagpur, Maharashtra, India
e-mail: shrikanttilloo@yahoo.com

S.S. Gurav and N.S. Gurav (eds.), *Indian Herbal Drug Microscopy*,
DOI 10.1007/978-1-4614-9515-4_3,

3. After the stain is taken up, wash off the excess of stain in water. Repeat the washing until stain stops coming out.
4. In some cases excess stain is removed by acid water or acid alcohol if water alone fails to do so.
5. Now the stained material is ready for mounting.

The following are few different stains and staining procedures which are primarily used for freehand sections only.

2.1 Toluidine Blue O

The stain TBO is an excellent stain for freehand sections. Being a polychromatic dye, TBO reacts with different chemical components of cells differently and results in a multicoloured specimen. The colours generated can provide information on the nature of the cell and its walls.

2.1.1 Stain Preparation

Dissolve 0.1 g of TBO in 100 mL of 0.1 M benzoate buffer at pH 4.4 (benzoic acid 0.25 g, sodium benzoate 0.29 g, water 200 mL). For general use, tap water can be used as the solvent for TBO, in case of unavailability of benzoate buffer.

2.1.2 Staining Procedures

1. Prepare sections as described in earlier chapter.
2. Select and place sections onto a clean slide.
3. Flood the sections with an aqueous solution of 0.1 % TBO solution for 1 min.
4. Gently remove the stain by using a piece of filter paper and wash the sections by flooding them with water followed by its removal. Repeat until there is no excess stain around the sections.
5. Add a drop of clean water over the sections and apply a cover glass. The slide is ready for examination.

2.1.3 Table 3.1: Results

Table 3.1 Results of TBO staining

Plant part/content	Colour after staining
Pectin	Red or reddish purple
Phenolic compounds	Green to blue green
Parenchyma and collenchyma	Reddish purple
Lignified elements and sclerenchyma	Green to blue green
Sieve tubes and companion cells	Purple
Middle lamella	Red to reddish purple
Callose and starch	Unstained

2.2 *Phloroglucinol–HCl*

Staining with phloroglucinol–HCl is most often an easy technique for colouring of lignin which is the common constituent in the secondary wall of plant cells. The cinnamaldehyde end groups of lignin react with phloroglucinol–HCl to give a red–violet colour.

2.2.1 Stain Preparation

Generally it is prepared as a saturated solution of phloroglucinol in 20 % hydrochloric acid (2 N). First dissolve phloroglucinol (about 2.0 g) in 80 mL of 20 % ethanol solution and then add 20 mL of concentrated HCl (12 N) to it.

2.2.2 Procedures

1. Prepare freehand sections as discussed in Chap. 1.
2. Place sections into a small petri dish and stain them with the phloroglucinol–HCl stain for two or more minutes.
3. Transfer sections onto a clean slide by using a wet brush and add a drop of water or a drop of 30 % glycerol solution to the section.

(Note: Do not forget to place a coverslip over the section before examination. The colour fades rapidly. Remember to wash the brush with running tap water to remove the acid.)

2.2.3 Table 3.2: Results

Table 3.2 Results of phloroglucinol–HCl staining

Plant part/content	Colour after staining
Heavy lignified walls	Deep red
Light lignified walls	Pale pink

2.3 *Iodine–Potassium Iodide*

The iodine–potassium iodide stain is specific for starch. Apparently, the basis of the reaction is the accumulation of iodine in the centre of the helical starch molecule. The length of the starch molecule determines the colour of the reaction, i.e., the shorter the molecule, the more red the colour; the longer the molecule, the more blue the colour.

2.3.1 Stain Preparation

First dissolve 2 g of KI in 100 mL of water. Then add 0.2 g of iodine into the KI solution. Prepare this solution ahead of time, as iodine takes some time to dissolve.

(Note: Store the solution in a tightly capped and dark glass bottle as exposure to light and air degrades the solution's usefulness.)

2.3.2 Procedure

1. Prepare freehand sections as described earlier.
2. Transfer sections onto a slide.
3. Place a drop of IKI solution directly on the specimen. Wait for a few minutes and apply a cover glass and examine the specimen with a microscope. The specimen can be examined without the removal of excess IKI solution from the sample.

2.3.3 Table 3.3: Results

Table 3.3 Results of iodine–potassium iodide staining

Plant part/content	Colour after staining
Starch	Blue–black colour
Newly formed starch	Red purple

2.4 Sudan Dyes

The mechanism of staining is based on differential solubility. As the Sudan dyes are more soluble in apolar solvents, they tend to dissolve more in hydrophobic structures such as the cuticle, lipid droplets or suberin.

2.4.1 Staining Solution

Dissolve 0.7 g of the Sudan IV in 100 mL of propylene or ethylene glycol. Heat the solution to 100 °C and stir it for several minutes. Filter the hot solution through Whatman No. 2 paper, cool and filter again.

2.4.2 Procedure

1. Prepare freehand sections as described earlier.
2. Transfer sections to petri dish containing Sudan IV staining solution and allow to stain for about 5 min.

3. Transfer the sections to 85 % propylene or ethylene glycol in water, and agitate the container gently for about 30 s to wash away excessive stain from the sections.
4. Briefly rinse the section with distilled water and mount in water or glycerol (glycerine; 30 % in water).

2.4.3 Table 3.4: Results

Table 3.4 Results of Sudan dye staining

Plant part/content	Colour after staining
Fats, oils and waxes	Red
Leaf cuticle, suberised walls and band	Red

2.5 Cytochemical Stains

Most of the stains are specific in reaction and purposely used so that definite structures or substances are stained. The following are few cytochemical stains used for staining different structures:

Achromatic figure: Aniline blue, erythrosine, fast green, light green
Cutinised cell wall: Crystal violet, erythrosine, safranin
Cellulose cell wall: Aniline blue, fast green, light green
Lignified cell wall: Crystal violet, safranin
Suberised cell wall: Safranin
Cytoplasm: Aniline blue, erythrosine, fast green, light green
Callose: Aniline blue
Chitin: Safranin
Proteins: Safranin
Mitochondria: Crystal violet
Plastids: Crystal violet, iron haematoxylin
Nucleus: Crystal violet, safranin, haematoxylin
Chromosomes: Haematoxylin, safranin

Some stains are used either alone or in combinations (commonly used) wherever tissue differentiation is necessary. A combination of acidic and basic dyes of contrasting colours is generally used in differentiation of woody tissue from nonwoody tissue.

3 Preparation of Powdered Drugs for Microscopic Observation

For microscopic observation, prepare three slides of powdered drug: one in chloral hydrate, one in water and one in phloroglucinol + conc. HCl.

Take a small quantity of powdered drug on a clean glass slide. Add two to three drops of chloral hydrate solution on it, and heat the slide very gently by passing to and fro over a low flame for 1–2 min. Care however is to be taken to avoid air bubbles and to see that there is sufficient chloral hydrate. Place a coverslip by holding it at an angle so no air bubbles creep in. Excess of chloral hydrate outside the coverslip is to be withdrawn using blotting paper. (Note: Chloral hydrate is used to clear the tissues and to bring in clarity of the view.)

Starch grains can be observed by mounting the powdered material in water. Just put one to two drops of dilute iodine solution in contact with the edge of the coverslip. Starch grains appear light blue in colour.

Add a few drops of mixture of 1:1 phloroglucinol + conc. HCl to the powder, cleared after chloral hydrate treatment as above. After 3–4 min, mount it in chloral hydrate or glycerine. One must take care to avoid the acid vapours from coming into contact with the microscopic objectives.

Different tissues can be confirmed by using different stains mentioned above.

Chapter 4
Herbal Drug Microscopy

Shailendra Gurav and Nilambari Gurav

1 Aconite (*Aconitum napellus*)

Synonyms:

Aconite root, Wolfsbane root, Monkshood, Bachnag, Fuzi, Monk's blood

Biological Source:

It consists of dried tuberous roots of the plant *Aconitum napellus* Linn.

S. Gurav (✉)
Government College of Pharmacy, Karad, Maharashtra, India
e-mail: shailugurav@rediffmail.com

N. Gurav
Sudhakarrao Naik Institute of Pharmacy, Pusad, Maharashtra, India
e-mail: nilagurav@rediffmail.com

S.S. Gurav and N.S. Gurav (eds.), *Indian Herbal Drug Microscopy*,
DOI 10.1007/978-1-4614-9515-4_4, © Springer Science+Business Media New York 2014

Family:

Ranunculaceae

Microscopy:

The transverse section of the tuberous root of aconite shows a wavy circular outer margin.

The following tissues are observed microscopically from the periphery towards the centre:

A. Metaderm:
   It is made up of tabular cells, which are thick walled, brownish in colour and irregularly arranged. It is also termed as 'epidermis' or 'epiblema'. This layer is not seen usually in the commercial samples of aconite. The walls of cells later become suberised and thus get converted into a protective tissue termed as 'metaderm'.
B. Primary cortex:
   This wide tissue is placed below the metaderm and is composed of about three to eight layers of tangentially elongated, thin-walled parenchymatous cells. Some of these cells are thick walled, lignified and pitted with large lumen. These rectangular cells are known as sclereids (stone cells) and generally seen as isolated or in small groups, interspersed within the cortical cells.
C. Endodermis:
   It is distinctly observed as a single layer of tangentially elongated, parenchymatous cells. Walls of the cells of endodermis are moderately thick and suberised.

D. Primary phloem:
   This area is wide and is composed of thick-walled, big parenchymatous cells. Starch granules are abundantly seen in these cells. A fine layer of pericycle and cambium is seen rarely above the primary phloem. It can be observed in the case of young root. Groups of sieve elements are seen as scattered in the primary phloem.
E. Parenchyma:
   This region also shows presence of small groups of sieve elements. This area is placed between the primary and secondary phloem.
F. Secondary phloem:
   This region mainly consists of parenchyma along with small scattered groups of sieve tissues. Smaller and numerous groups of sieve elements are observed near the cambium.
G. Cambium:
   It is made up of two to three rows of small cells. It appears as a stellate ringlike band.
H. Xylem:
   It is just below the cambium. It shows small groups of pitted and reticulate vessels embedded irregularly in the parenchyma. Radially elongated small groups of xylem vessels are also seen near the pith.
I. Pith:
   It is narrow and made up of radially elongated to isodiametric parenchymatous cells which abundantly contain starch.

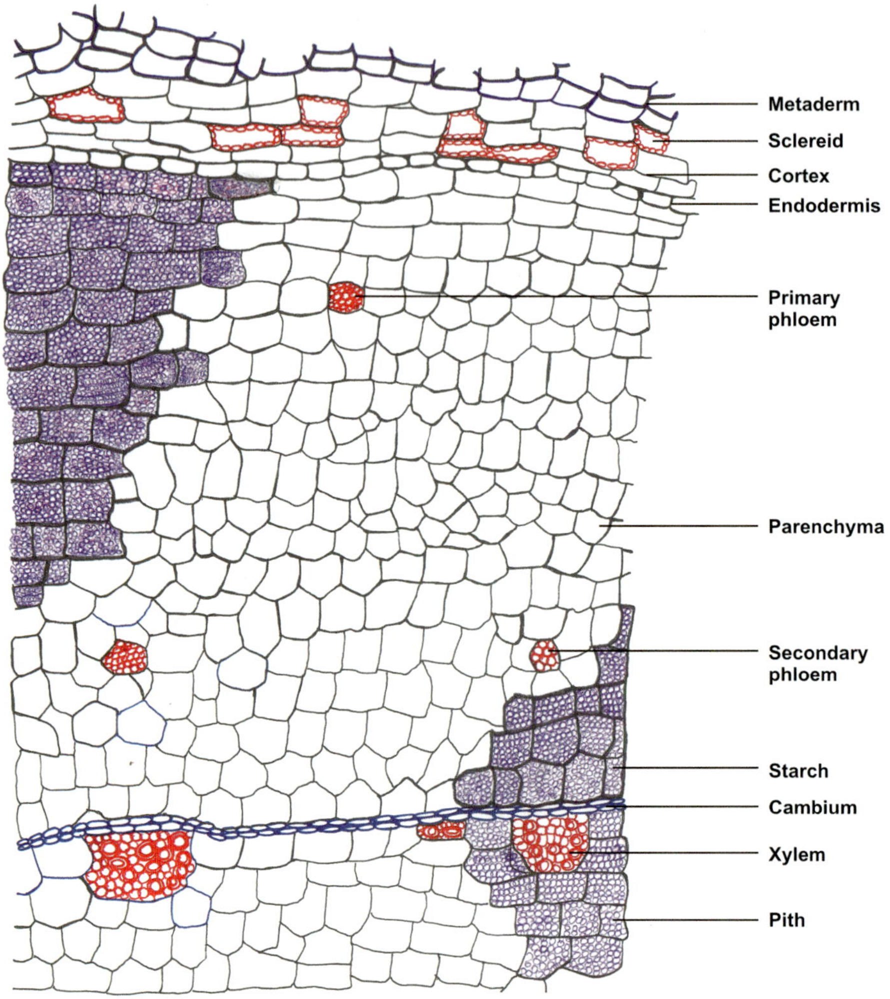

T. S. of Aconite root

Powder Character:

Powdered aconite root is greyish brown in colour with faint odour and sweetish taste, followed by persistent sensation of numbness. The powder shows the following diagnostic characters:

a. Starch granules:
   The abundant starch granules are simple and spherical, sometimes compound with two, four or up to six components. Larger granules show a radiate or slit-shaped hilum.
b. Sclereids:
   These are numerous and occur singly with associated thin-walled parenchymatous cells in small groups. These are large with oval, square or sub-rectangular outline, pitted thick walls and a large lumen.
c. Parenchyma:
   The parenchyma of the cortex and pith is abundantly observed with large cells which vary from round to elongated in outline; walls are thick with pits. Cells are filled with starch granules.
d. Fragments of outer layer:
   These appear dark brown to black in colour. In surface view, cells appear sub-rectangular, thick walled and unevenly pigmented.
e. Vessels:
   These are large, single or in small groups. Walls are lignified and show numerous slit-shaped pits with indistinct borders. Few vessels occur with reticulate, spiral or annular thickening.
f. Fibres:
   Fibres from stem bases occur occasionally which are lignified and thin walled with numerous pits.

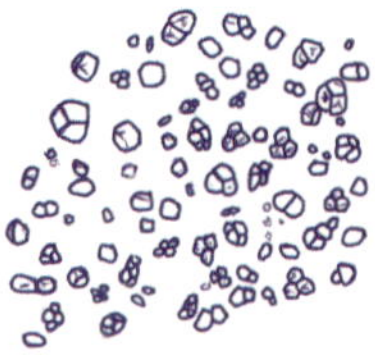

Starch granules

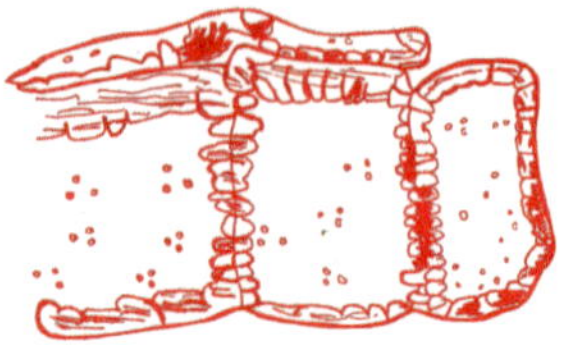

Part of a group of sclereids

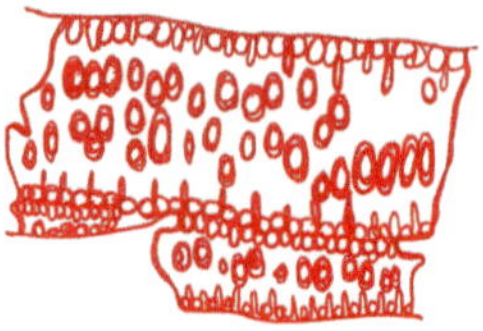

Fragments of
bordered pitted vessels

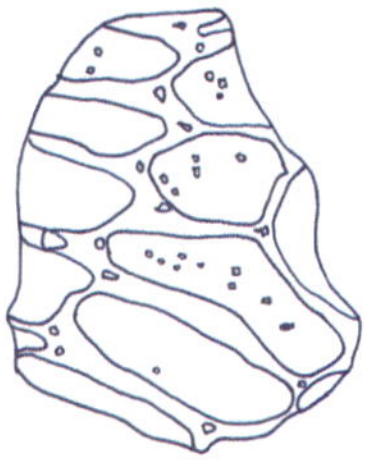

Thick-walled parenchyma

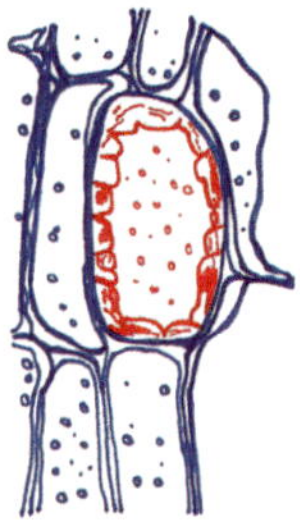

A single sclereid with associated
thicker-walled parenchyma

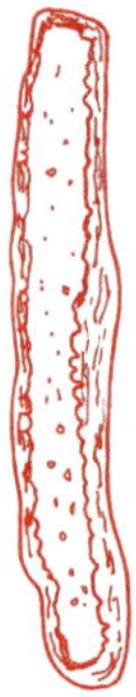

Fibrous sclereid
from the stem base

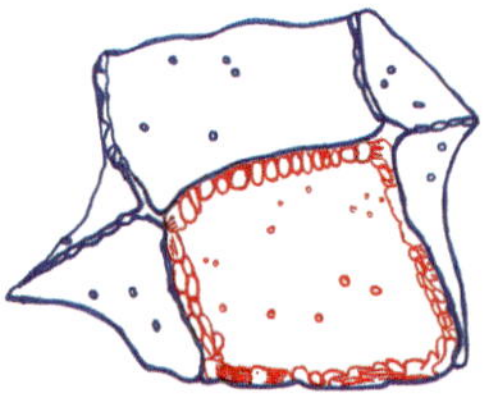

A single sclereid with associated
thin-walled parenchyma

Part of a fibre
from the stem base

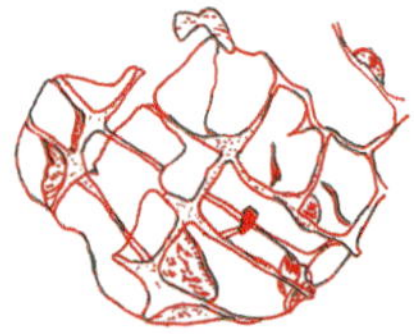

Outer layer in surface
view showing pigment

2 Aloe (*Aloe barbadensis*)

Synonyms:

Aloe, Musabbar, Kumari, Korphad

Biological Source:

It consists of dried juice of the leaves of the plant *Aloe barbadensis* Miller, known as Curacao aloes, or of *Aloe perryi* Baker, known as Socotrine aloes, or of *Aloe ferox* Miller and hybrids of this species with *Aloe africana* Miller and *Aloe spicata* Baker, known as Cape aloes.

Family:

Liliaceae

Microscopy:

Microscopically the transverse section of a leaf of aloe shows the following zones:

A. Epidermis:
   It consists of a single row of more or less circular cells with thick cuticle. Epidermis of both surfaces shows numerous stomata.
B. Palisade:
   A single row of rectangular palisade cells can be seen.
C. Parenchyma:
   Below the palisade, a zone of parenchyma is seen. It shows chlorophyll, starch and bundles of needles of calcium oxalate in some cells.

D. Mucilaginous parenchyma:
This central region occupies the major area (nearly three fifth) of the leaf diameter and gives an aloe leaf a fleshy appearance. Many large mucilage-containing parenchymatous cells are observed.

E. Vascular bundles:
Vascular bundles are seen in double row near the parenchymatous zone. These bundles are adjacent to the pericycle and endodermis. Pericyclic cells and parenchyma yield the aloetic juice.

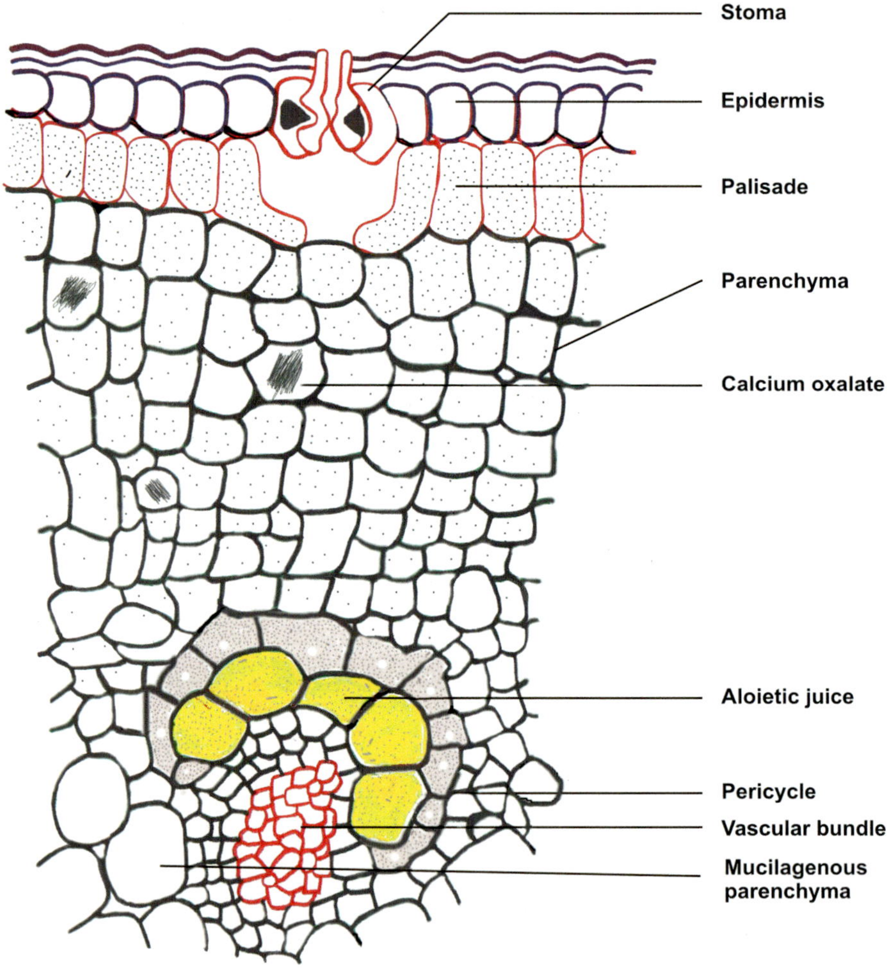

T. S. of Aloe leaf

3 Anise (*Pimpinella anisum*)

Synonyms:

Anise fruit, Aniseed, Badiyan rumi

Biological Source:

It consists of dried ripe fruits of *Pimpinella anisum* Linn.

Family:

Umbelliferae

Microscopy:

Anise shows features of a typical umbelliferous fruit.

Cremocarp:
Cremocarp is a variety of schizocarp (splitting fruit) which divides into two one-seeded portions, each corresponding to one carpel. This carpel itself does not open to liberate the seed; thus, these schizocarps are indehiscent fruits. A cremocarp consists of two parts, each of which is called a 'mericarp'. These two mericarps are connected by a thick-walled sclerenchymatous central stalk called 'carpophore'. A single seed is seen in each mericarp. Raphe is a single ridge of vascular bundle at the middle of the commissural surface. The carpophore is situated just in front of the raphe.

Transverse section of a mericarp shows two prominent surfaces: commissural and dorsal. The commissural surface is flat with two distinct ridges and carpophore in the middle. The dorsal surface shows three ridges. Therefore, each mericarp shows five primary ridges.

A mericarp can be divided into pericarp and seed.

A. Pericarp:
   It is made up of epicarp, mesocarp and endocarp:

   i. Epicarp (exocarp):
      The epicarp of the pericarp is also called epidermis. It consists of a layer of thick-walled, polygonal, tabular, tangentially elongated cells with striated cuticle bearing covering trichomes and frequent stomata. Covering trichomes are unicellular, thick walled, short, conical, warty and slightly curved and are usually found detached from the epicarp.
   ii. Mesocarp:
      The bulk of the mesocarp is parenchymatous. It contains much thickened and lignified parenchyma in the region of the vascular strands. These thickened walls have large, oval or rounded pits. These bands of thickening between them give a reticulate appearance to the walls. Mesocarp shows five vascular bundles below each ridge (thus ridges are known as primary ridges). Reticulate lignified parenchyma is seen around the vascular bundles. Vascular bundles are five in number, bicollateral and present below each ridge. The remaining parenchyma of the mesocarp is made up of small, polyhedral, cellulosic cells. Mesocarp shows presence of groups of sclereids associated with thin-walled, unlignified parenchyma. Isolated sclereid has a square to rectangular margin and a large lumen. It shows uniformly thickened walls transversed by numerous pits. The mesocarp of each mericarp shows two (sometimes three or four) large vittae on the commissural surface and about twenty to forty small vittae (arranged in a row) on the dorsal surface. Each vitta is lined by an epithelium of small polygonal tabular cells.
   iii. Endocarp:
      It consists of narrow elongated, thin-walled cells having a parquetry arrangement. In the transverse section, long narrow rectangular cells are seen along with scattered groups of very short cells.

B. Seed:
   It consists of testa, endosperm and embryo:

   i. Testa
      It is single layered and yellowish brown in colour. It is made up of polygonal cells with thin and beaded walls.
   ii. Endosperm:
      It is composed of thick-walled, polygonal, colourless, cellulosic parenchyma containing fixed oil globules, aleurone grains and micro-rosette crystals of calcium oxalate.

iii. Embryo:
A crescent-shaped embryo is observed in the sections passing through the apical region of mesocarp. It is seen at the centre of the endosperm. Cells are polygonal, thick walled and parenchymatous.

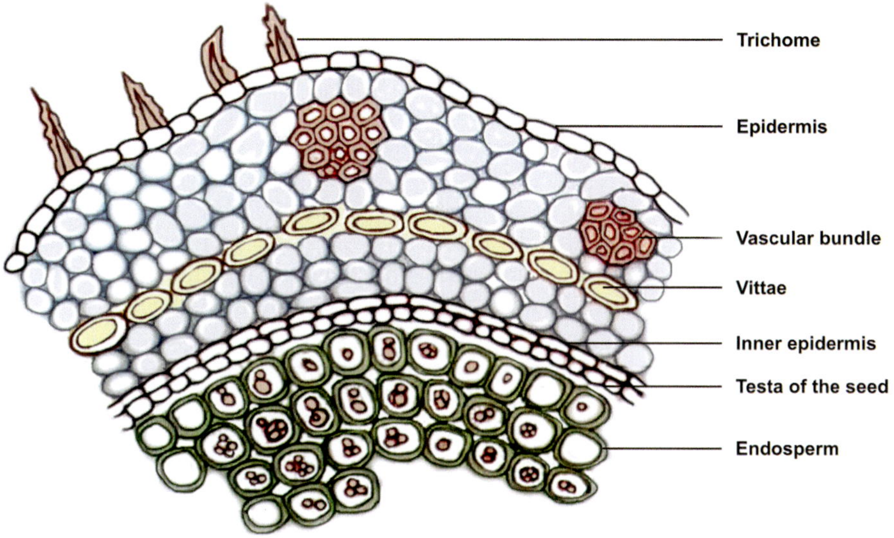

T. S. of Anise fruit

Powder Character:

The powder of anise is yellowish brown with characteristic, aromatic odour and taste. It shows the following characters microscopically:

a. Epicarp:
It is composed of a layer of colourless, thin-walled cells, polygonal in surface view and with a strongly striated cuticle. It may show few scattered stomata. Stomata may be surrounded by two to four radiating cells and are found on some fragments of epicarp near the sinuous walls. Fragments of epicarp may show slight thickening and beading of the straight walls.
b. Endocarp:
It consists of a layer of very thin-walled, lignified cells, elongated in surface view and arranged in groups of about six or more cells with their long axes parallel to each other, i.e. parquetry arrangement of cells. This layer is generally found attached to fragments of the vittae. Larger fragments of endocarp show a row of small vittae as yellow band or zone of oil cells.
c. Vittae:
Numerous brown fragments of vittae are observed composed of thin-walled, polygonal to elongated cells with several branching septa. These appear irregular in shape and found scattered or attached to the pieces of parquetry layer of endocarp. Larger fragments may exhibit branching of the vittae.

d. Endosperm:
   Fragments of endosperm are abundant and are composed of polygonal, thick-walled cells with fixed oil globules and aleurone grains. These cells also show presence of micro-rosette crystals of calcium oxalate.
e. Stone cells (sclereids) of the mesocarp:
   A few sclereids occur in groups in a single layer and attached to thin-walled unlignified parenchymatous cells of mesocarp. A single sclereid is rectangular with large lumen and uniformly thick pitted walls. The innermost layer of mesocarp consists of slightly thick-walled cells, rounded to rectangular in surface view. This layer is found associated with the endocarp.
f. Fibrovascular tissue:
   Fragments of fibrovascular tissue are found scattered. These are composed of lignified, small, thin-walled fibres and vessels with spiral and annular thickening. Some of these fragments are associated with reticulate parenchyma of the mesocarp. A few fragments of fibrovascular tissues are also found scattered.
g. Testa:
   It is composed of a single layer of brownish cells with thin and slightly beaded walls.
h. Trichomes:
   The covering trichomes are found as detached from the epicarp. These are unicellular, thick and warty walled, conical and slightly curved. Occasionally the lumen is divided by a transverse septum.

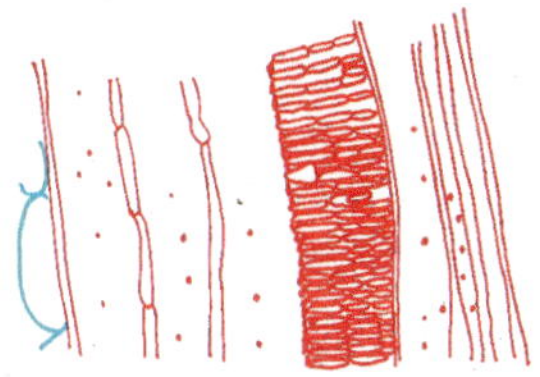

Part of a group of
fibro-vascular tissue

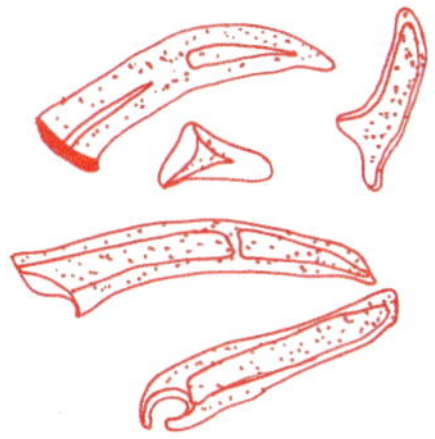

Covering trichome

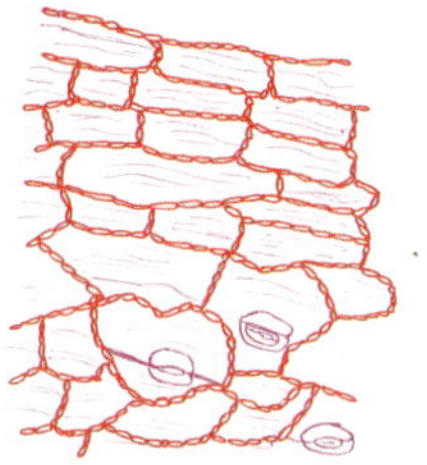

Epicarp in surface view
showing stomata and
striated cuticle

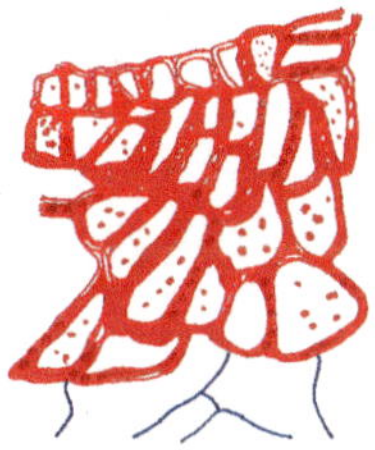

A group of sclereids
from the mesocarp with
adjacent unlignified parenchyma

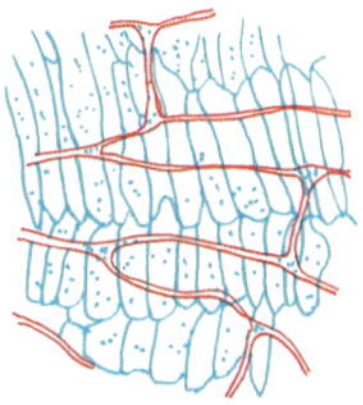

Branching vittae
(showed in outline only)
and underlying
endocarp in surface view

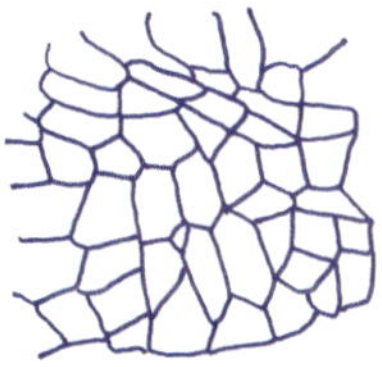

Testa in surface view

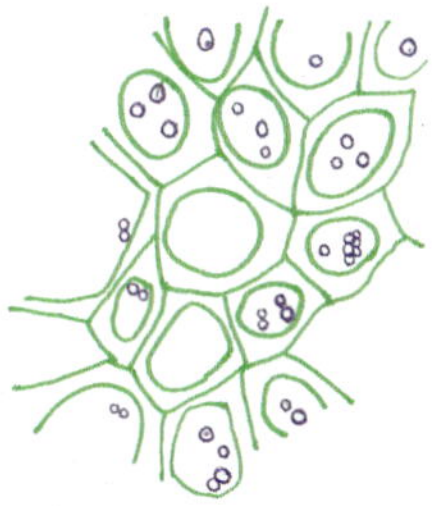

Endosperm containing
microrossette crystals
of calcium oxalate

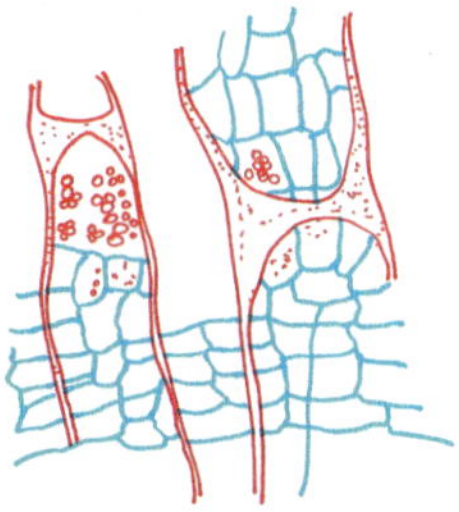

Part of two vittae
showing transverse septa
and part of the underlying
endocarp in surface view

4 Arjuna (*Terminalia arjuna*)

Synonyms:

Arjuna, Dhavala, Arjana, *Arjun*, *Marutha maram*, *Terminalia glabra* W and A

Biological Source:

It consists of dried stem barks of the plant *Terminalia arjuna* (Roxb.) Wight. and Arn.

Family:

Combretaceae

Microscopy:

The transverse section of the arjuna bark shows the following tissues microscopically:

A. Cork:
   It is composed of many uniformly arranged layers of small, tangentially elongated cells.

B. Cortex:
It is a broad zone below the cork consisting of thin-walled, brick-shaped, rectangular parenchymatous cells containing cluster crystals of calcium oxalate. A few groups of sclerenchymatous pericyclic fibres are found scattered in the cortex.

C. Secondary phloem:
It consists of phloem parenchyma whose cells are polygonal with thin and wavy walls. It shows cluster crystals of calcium oxalate and pigmented cells. Phloem fibres are composed of sclerenchymatous cells and occur in groups and are also found scattered in the form of patches. Young stem bark also shows mucilage, secreting ducts, sclerenchyma of fibres and tanniferous cells. Mature bark shows a broad zone of phloem consisting of ceratenchyma, phloem parenchyma, phloem fibres and crystals fibres. The crystal fibres contain rosette crystals of calcium oxalate.

D. Medullary rays:
These are narrow, many layered and almost straight. These rays are radially elongated and parenchymatous with small pits and starch grains. A few cluster crystals of calcium oxalate may appear.

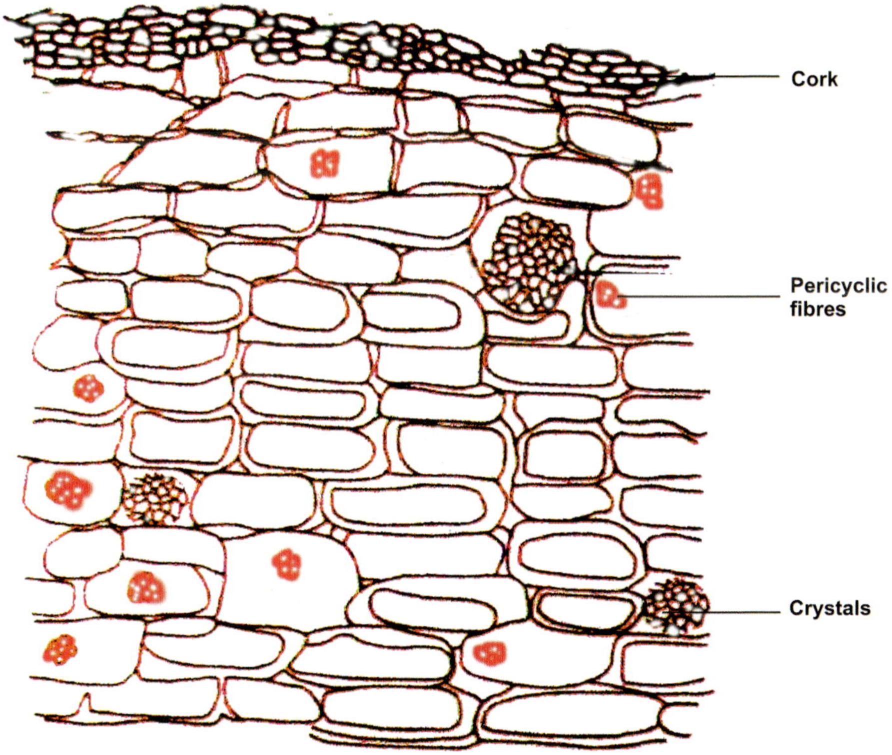

T. S. of outer part of the bark

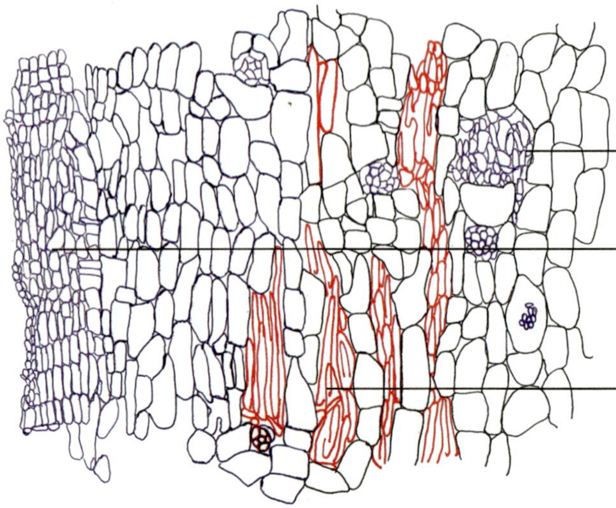

L. S. of outer part of the bark

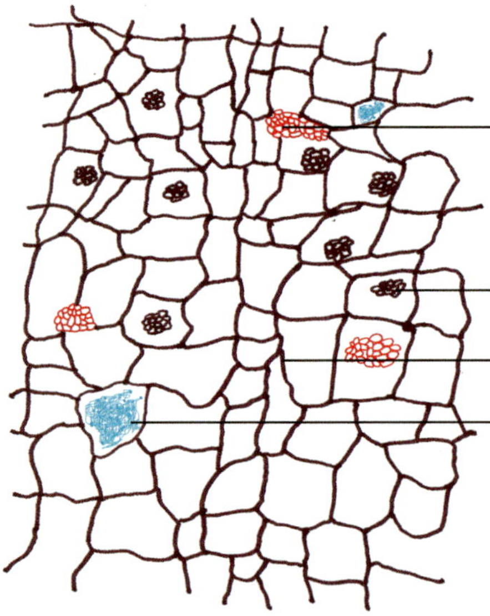

T. S. of inner part of bark

Powder Character:

The powder of arjuna bark appears pinkish or reddish brown with slight odour and astringent taste. It shows the following characters microscopically:

a. Cork:
   These cells are seen as with moderately thick walls and polygonal in surface view. If the bark is exfoliated as it is in commercial samples, only few fragments of cork cells are observed.
b. Starch granules:
   These are present in various tissues like cork, cortex and medullary rays and thus occur abundantly. Simple or compound grains (with two to seven components) occur and show a distinct central hilum.
c. Calcium oxalate crystals:
   Cluster crystals of calcium oxalate are observed. A few are large in size. These are more in number in the case of matured stem bark powder.
d. Parenchymatous cells:
   Fragments of parenchymatous cells are observed. These are thin walled and rectangular in shape. Some cells contain cluster crystals of calcium oxalate. Parenchymatous cells of phloem are polygonal with thin and wavy cell walls. A few of these cells also show cluster crystals of calcium oxalate.
e. Medullary rays:
   These are multiseriate and made up of lignified parenchymatous cells. Fragments are found attached to the cells of secondary phloem.
f. Fibres:
   Phloem fibres made up of sclerenchymatous cells appear as scattered. Sclerenchymatous pericyclic fibres with tapering ends are also observed.
g. Pigment matter and other contents:
   Isolated masses of pigment appear as scattered in the powder. Mucilage cells, tanniferous cells and secreting ducts may be seen.

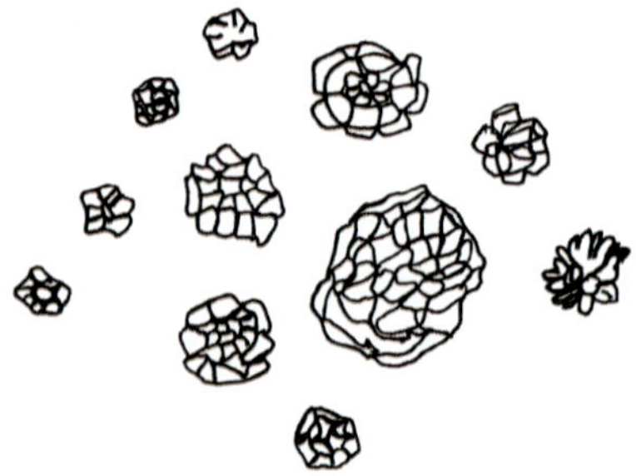

Calcium oxalate crystals

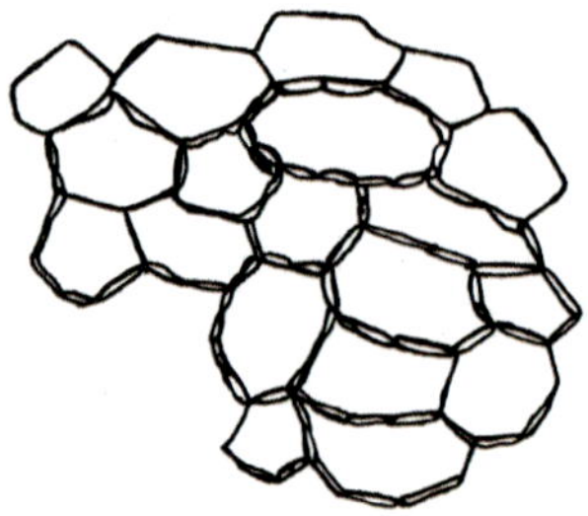

Cork in surface view

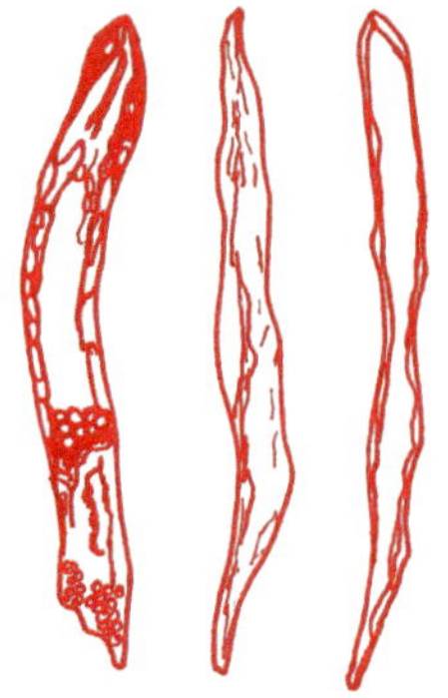

Fibres

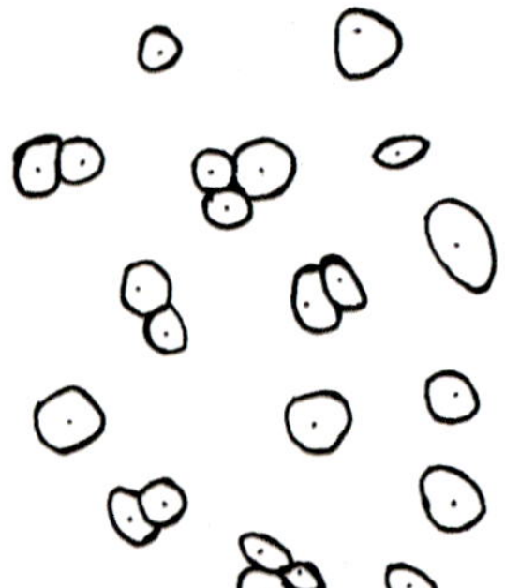

Starch granules

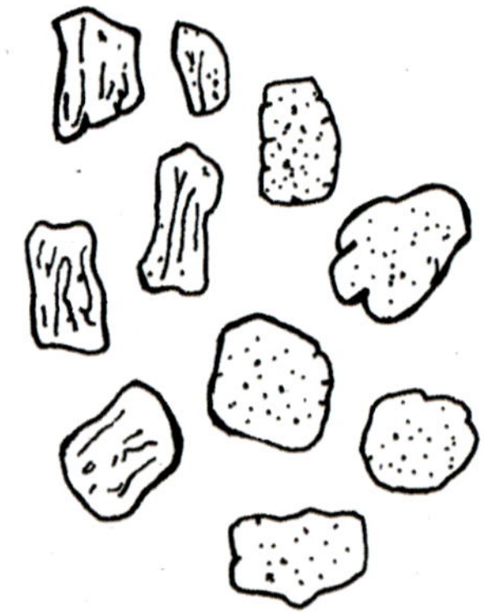

Isolated masses of pigment

5 Ashoka (*Saraca indica*)

Synonyms:

Asoka, Ashoka, Ashoka bark, Hempushpa, *Saraca asoca* (Roxb.) DC Wilde

Biological Source:

It consists of dried stem bark of the plant *Saraca indica* Linn.

Family:

Leguminosae

Microscopy:

Transverse section of the ashoka bark shows periderm, pericycle and wide secondary phloem. The following different tissues are observed from the periphery towards the centre:

A. Periderm:
   It is made up of about ten to 25 layers of cork (phellem) cells, three to four layers of phellogen and a wide phelloderm. Cork cells appear polygonal in surface view and possess suberised walls.

B. Pericycle:
It is made up of sclereids, parenchyma and pericyclic fibres:

i. Sclereids:
These are seen as densely packed zone which is made up of thick-walled, tangentially elongated cells. These sclereids are placed alternate with the parenchyma.
ii. Parenchyma:
Parenchymatous cells are thick walled and oval shaped containing prisms of calcium oxalate. These cells are seen alternate to the sclereids. A sheath of prisms of calcium oxalate encircles the sclereids.
iii. Pericyclic fibres:
Pericyclic fibres are found as scattered. These fibres show straight margins, tapering ends and narrow lumen.

C. Secondary phloem:
It is wide and made up of phloem parenchyma, phloem fibres and medullary rays:

i. Phloem parenchyma:
It is similar to the parenchyma of the pericycle. Prisms of calcium oxalate crystals are observed inside cells of phloem parenchyma.
ii. Phloem fibres:
These are regularly arranged in small concentric groups of three or more fibres on radial rows of the phloem elements. Sieve tubes are clearly visible.
iii. Medullary rays:
Medullary rays vary in width in outer and middle regions and are dilated at their outer ends. These rays are seen as funnel shaped near the pericycle. This part of the bark is characteristic. The wider, funnel-shaped medullary rays transversely cut the narrow radial rows of the phloem. Thus, sieve tubes and companion cells are crushed mostly and form 'ceratenchyma'. Some of the medullary ray cells become thick walled and lignified and get converted into stone cells (sclereids).

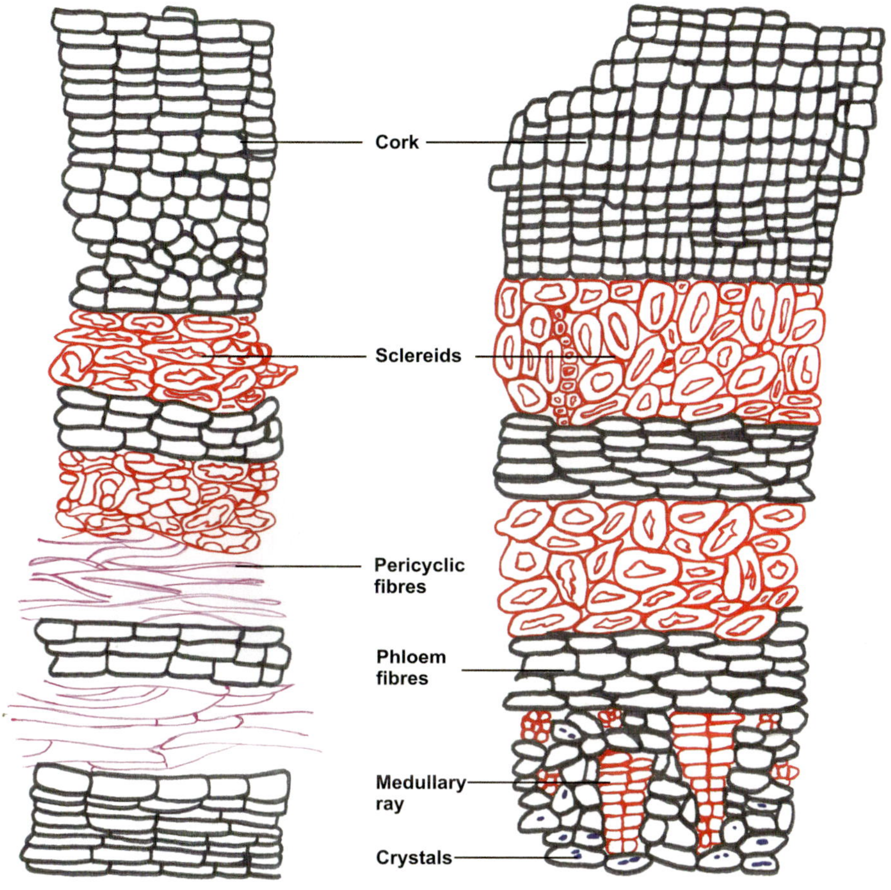

L. S. of Ashoka bark

T. S. of Ashoka bark

Powder Character:

The powder of ashoka bark is rusty brown in colour with astringent taste. It shows the following characters microscopically:

a. Sclereids:
   Groups of sclereids are found as scattered in the powder. These are composed of compactly placed cells which are thick walled and heavily pitted. These are found as isolated as well as associated with the sheath of prismatic crystals of calcium oxalate. Sometimes sclereids show presence of a yellowish brown matter.
b. Cork cells:
   These are moderately thick walled and polygonal in surface view.
c. Phloem fibres:
   These are generally seen as grouped lengthwise in groups of three to five fibres. In each group occasionally these fibres are found as scattered and show tapering ends and some brownish matter attached to them.
d. Calcium oxalate crystals:
   These are small prismatic crystals of uniform size and found scattered. These are contained in the parenchymatous cells as well as placed as sheath of small prisms surrounding the sclereids.
e. Starch grains:
   Small, simple, spherical starch granules are seen occasionally.

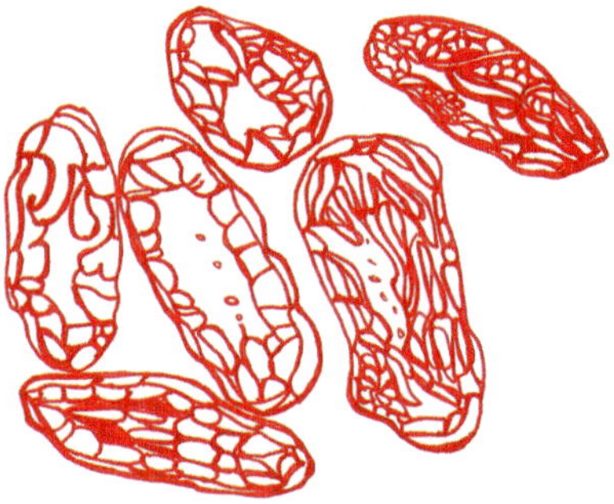

Sclereids

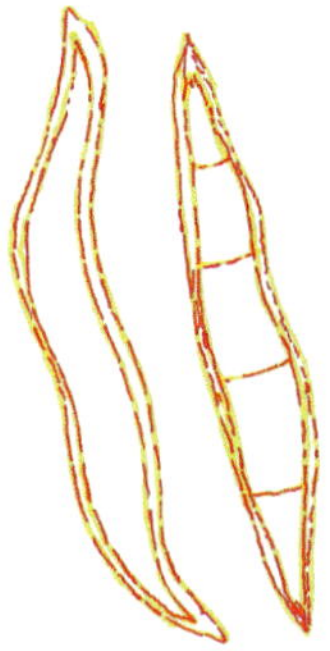

Fibres

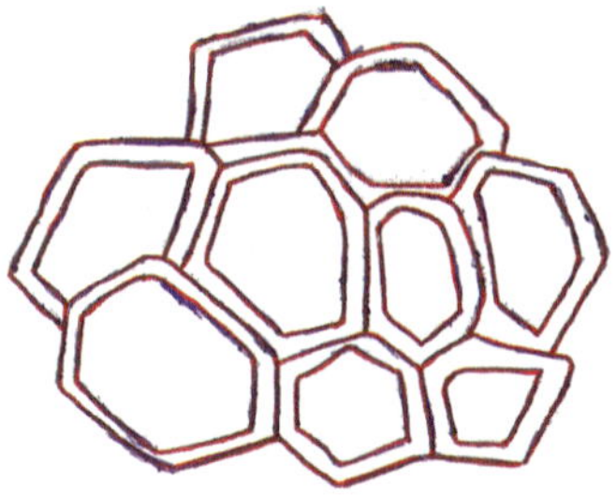

Cork in surface view

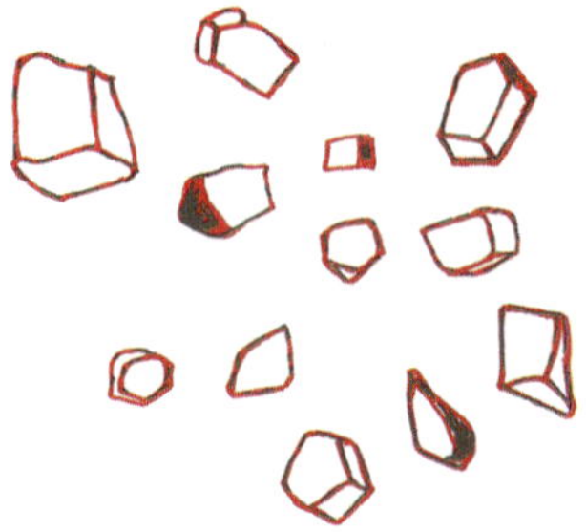

Calcium oxalate crystals

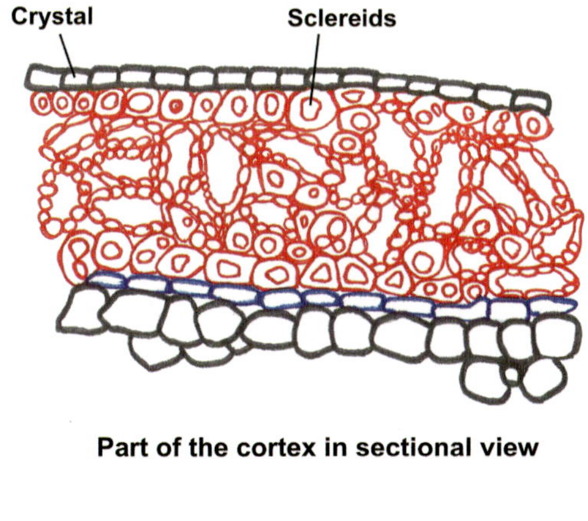

Part of the cortex in sectional view

6 Ashwagandha (*Withania somnifera*)

Synonyms:

Withania root, Asgandh

Biological Source:

It consists of dried roots of the plant *Withania somnifera* Linn.

Family:

Solanaceae

Microscopy:

The transverse section of the ashwagandha root shows the following tissues:

A. Cork:
   It consists of several (two to eight) layers of isodiametric, non-lignified, thin-walled, suberised, uniformly placed parenchymatous cells. Cork cambium is single layered or sometimes indistinct.

B. Cortex:
   It is broad and occupies major portion of the root. It is composed of thin-walled, polygonal, flattened, irregular parenchymatous cells containing starch grains.
C. Phloem:
   It is made up of isodiametric parenchymatous cells with intercellular spaces. Phloem of old root shows sieve tubes, companion cells and parenchyma.
D. Secondary xylem:
   It is present below parenchymatous cambium. It consists of xylem parenchyma, lignified vessels, tracheids, fibres and medullary rays.
E. Medullary rays:
   These are single celled, multiseriate and straight. The longitudinal section shows vessels and tracheids with pitted thickening.
F. Ground tissue:
   It is parenchymatous, present below medullary rays and filled with starch grains.
G. Primary xylem:
   It is seen sometimes in the central area.
   In the case of old roots, cork is exfoliated or crushed. Phelloderm shows many layers of compact parenchymatous cells. Phloem exhibits sieve tubes, companion cells and phloem parenchyma. Xylem contains hard and vascular rings close to each other, separated by multiseriate medullary rays.

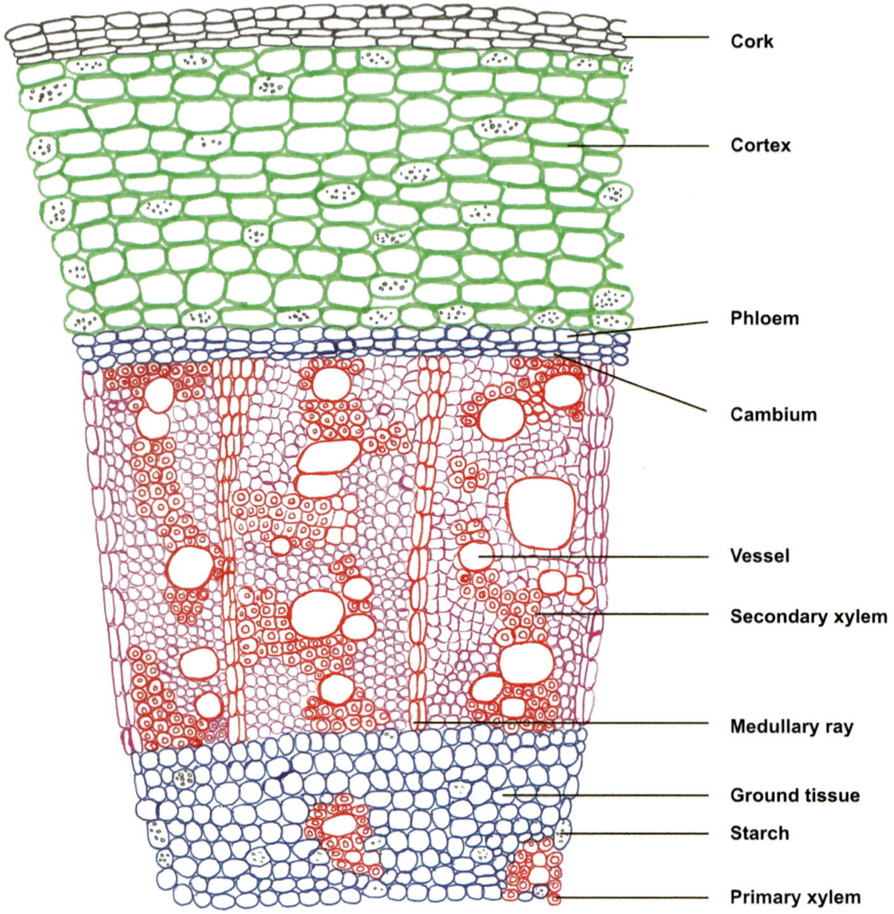

T. S. of Ashwagandha root

Powder Character:

The powder of ashwagandha root is greyish with characteristic odour and bitter, acrid taste. It shows the following characters microscopically:

a. Cork:
   Cork cells have an irregular outline and are thin walled and polygonal.
b. Vessels:
   These are abundantly present and lignified with bordered pits and spiral and annular thickening. Tracheids have pitted thickening.
c. Fibres:
   These are narrow, lignified, thick walled and elongated with tapered ends.
d. Calcium oxalate crystals:
   Microsphenoidal crystals of calcium oxalate are seen.
e. Starch grains:
   These are mostly simple with a few compound grains and spherical-ovoid with distinct hilum at the centre. These are found scattered or abundant in parenchymatous cells of the cortex.

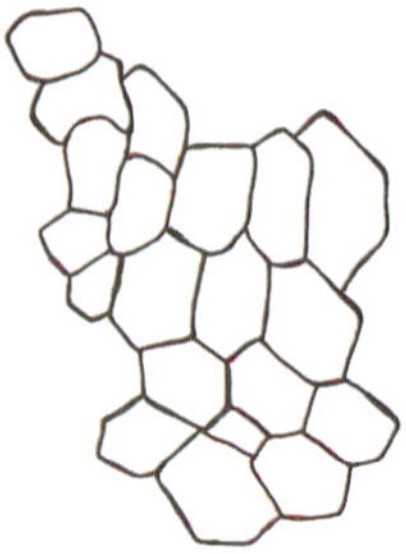

Cork cells
in surface view

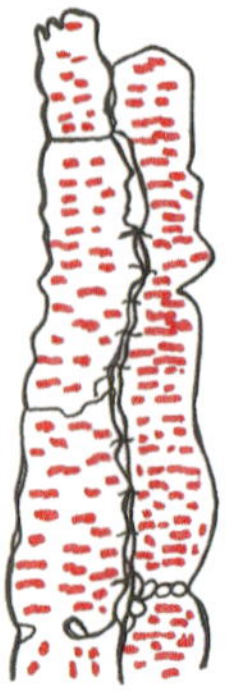

Tracheids

Fibres

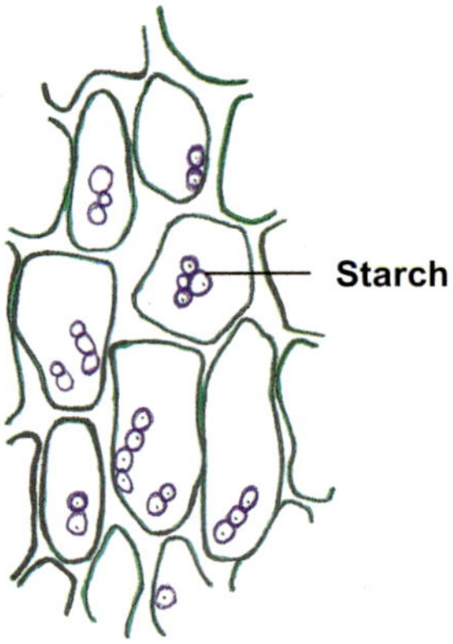

Part of cortex in sectional view

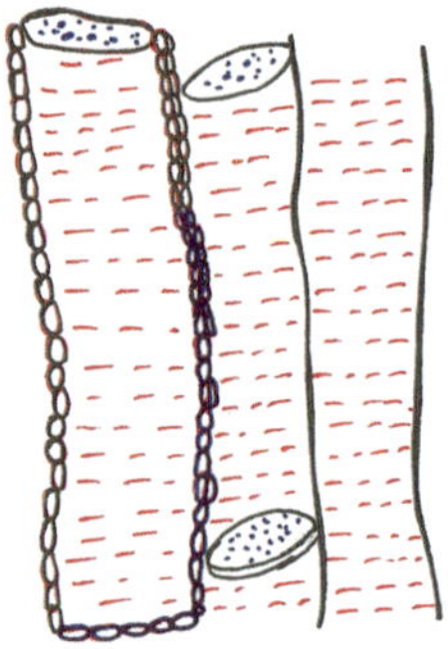

Vessels

Starch granules

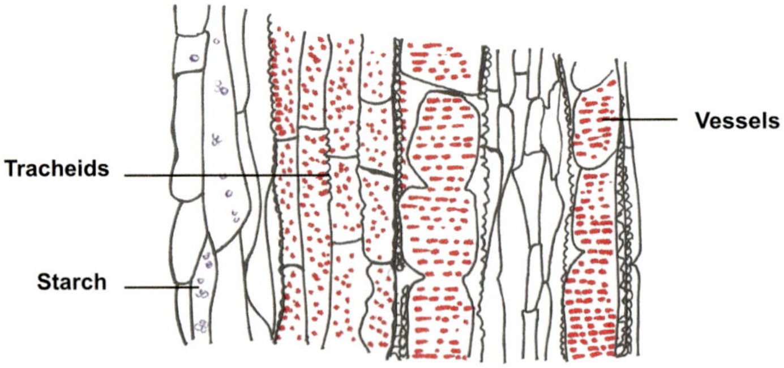

L. S. through xylem portion

7 Asparagus (*Asparagus racemosus*)

Synonyms:

Shatmuli, Shatavari, Satavar

Biological Source:

It consists of dried roots of the plant *Asparagus racemosus* Wild.

Family:

Liliaceae

Microscopy:

The transverse section of the asparagus root shows a typical monocot structure. Microscopically it shows the following characters:

A. Epidermis:
   It is single layered with tangentially elongated, thin-walled cells with a large number of thin-walled hair rootlets.

B. Exodermis:
   Below the epidermis, there are three to four layers of suberised thick-walled cells.
C. Cortex:
   Cortex occupies the major part of the transverse section. It is composed of thin-walled, spherical-ovoid parenchymatous cells containing bundles of acicular crystals of calcium oxalate, mucilage and a few starch granules.
D. Endodermis:
   Below the cortex, there are three to four layers of lignified, thin-walled cells.
E. Pericycle:
   It is composed of a single layer of non-lignified thin-walled cells.
F. Vascular bundles:
   The central stele shows radially arranged vascular bundles below the endodermis. The vascular bundle consists of alternate strands of xylem and phloem. Phloem is non-lignified and appears like a bunch of grapes. Xylem is exarch in nature and lignified. It exhibits xylem parenchyma and xylem vessels. Vessels have spiral thickening and tracheids have pitted thickening.
G. Pith:
   These cells are parenchymatous and lignified as well as some non-lignified with intercellular spaces.

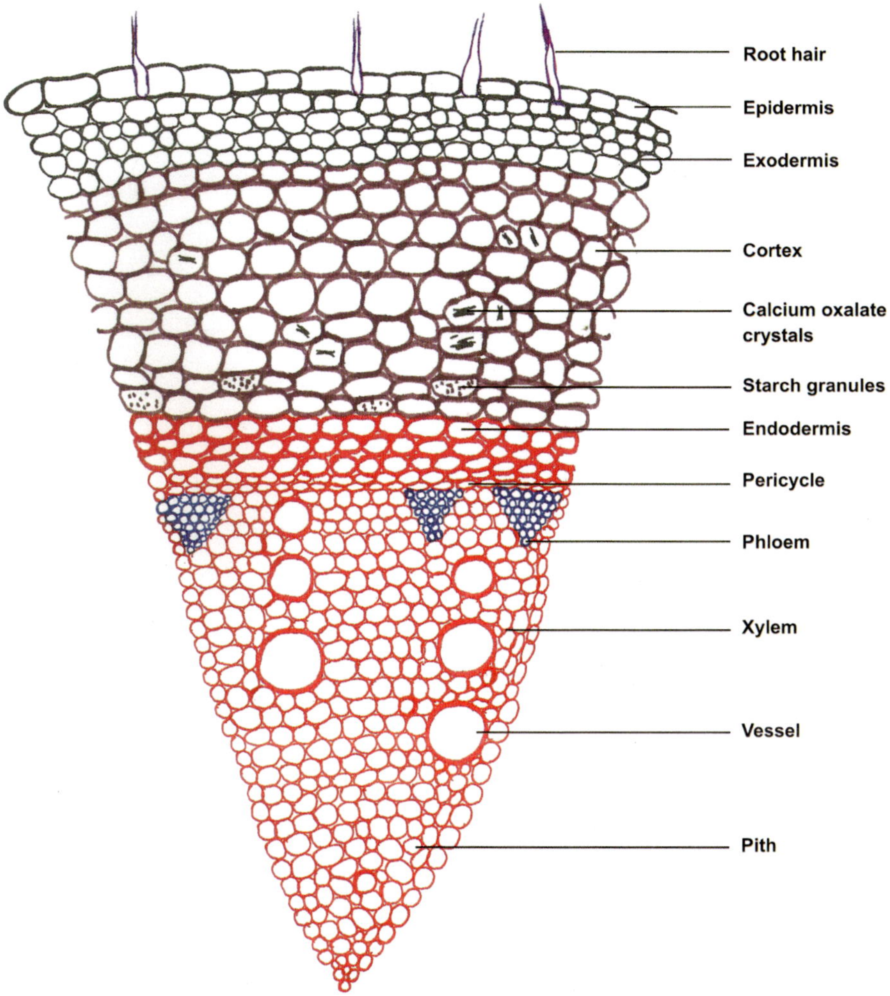

T. S. of Asparagus root

Powder Character:

The powder of asparagus root is whitish with slight odour and taste. Microscopic examination shows the following characters:

a. Epidermal cells:
   These are thin walled with hair rootlets.
b. Xylem vessels:
   These are found as abundantly as lignified fragments. These have spiral thickening and tracheids have pitted thickening.
c. Parenchyma:
   Thin-walled, lignified parenchymatous cells of the cortex are found and a few pitted parenchyma of pith are found in the fragments.
d. Calcium oxalate crystals:
   Acicular crystals of calcium oxalate are found scattered or usually in bundles inside the parenchymatous cells of the cortex.

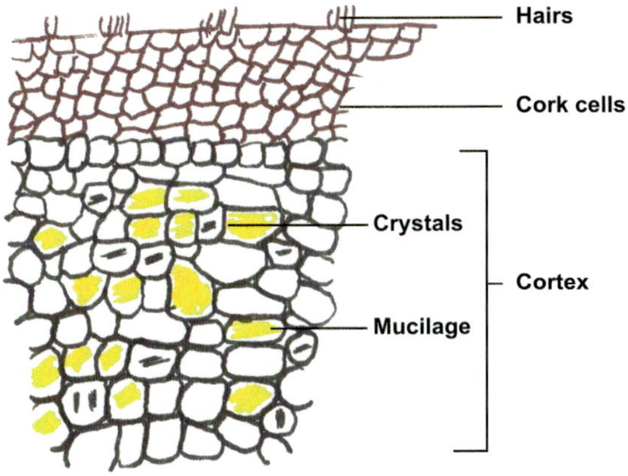

T. S. of outer part of root

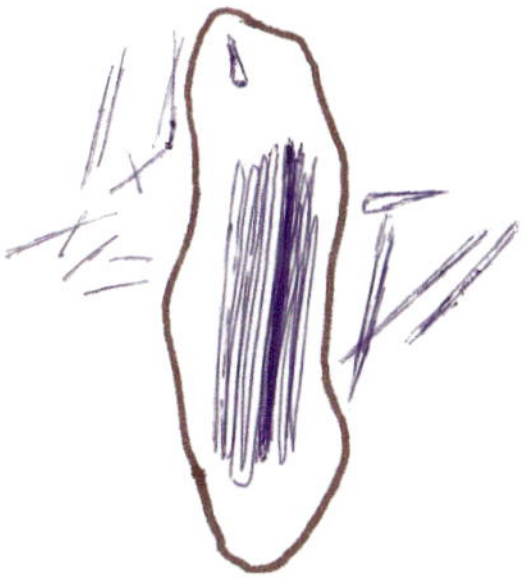

Acicular calcium oxalate crystals

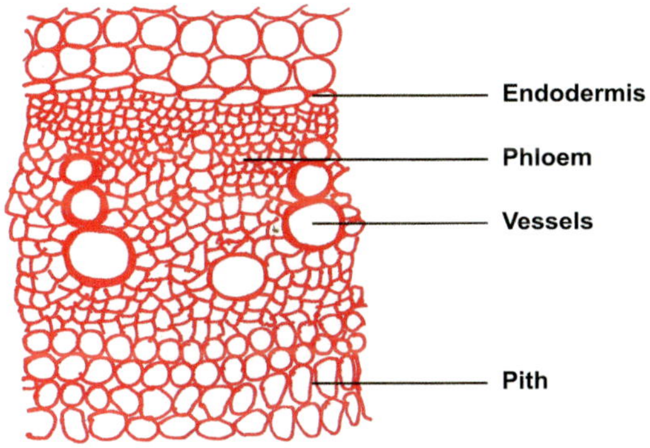

T. S. through xylem portion

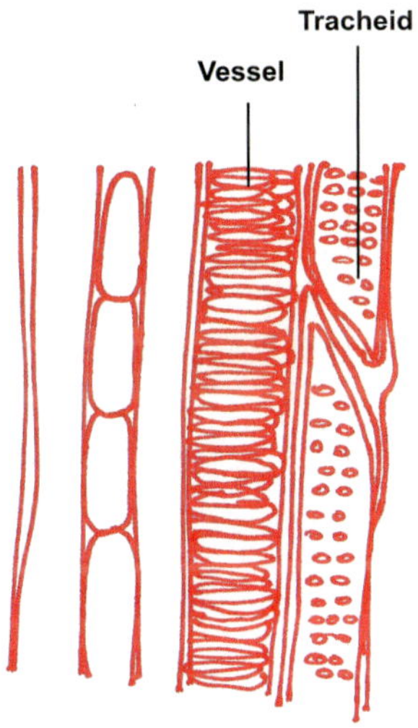

L. S. through xylem portion

8 Belladonna (*Atropa belladonna*)

Synonyms:

Belladonna leaf, Belladonna herb, Deadly nightshade leaf

Biological Source:

It consists of fresh or dried leaves and flowering tops of *Atropa belladonna* Linn. (European belladonna) and *Atropa acuminata* Royle. Ex. Lindl. (Indian belladonna).

Family:

Solanaceae

Microscopy:

The transverse section of the young belladonna root appears as a ring with wavy circular outer margin. It shows the following tissues from the periphery towards the centre:

A. Periderm:
   It can be differentiated into phellem and phellogen:

   i. Phellem (cork):
      The cork is made up of few layers of tangentially elongated cells which are arranged radially.
   ii. Phellogen:
      A very thin layer exhibits phellogen and is not distinctly seen usually.

B. Phelloderm:
   It is composed of a few layers of tangentially elongated cells, some of which contain starch granules and yellowish masses. Occasionally a few cells show presence of fine crystalline sand of calcium oxalate. Small groups of fine sand of calcium oxalate are formed and known as 'sandy balls'.
C. Secondary phloem:
   This region is made up of many layers of parenchymatous cells along with scattered groups of sieve elements. These parenchymatous cells show presence of abundant starch within them. Sandy balls are found scattered throughout the phloem tissue, and this is a characteristic feature of belladonna root and leaf.
D. Cambium:
   It is in the form of a ring of four to five rows of thin-walled rectangular cells.
E. Secondary xylem:
   It forms the major portion of the root. It mainly consists of parenchyma, vessels, tracheids, fibres and cellulosic parenchyma. Parenchymatous cells show presence of starch abundantly within them. Xylem vessels are in the form of groups of three to ten and found as scattered. These groups of vessels are more crowded near the cambium. Vessels are associated with a few tracheids, fibres and cellulosic parenchyma. Numerous characteristic sandy balls are found all over the secondary xylem. A distinct diarch structure of primary phloem is also observed at the centre.

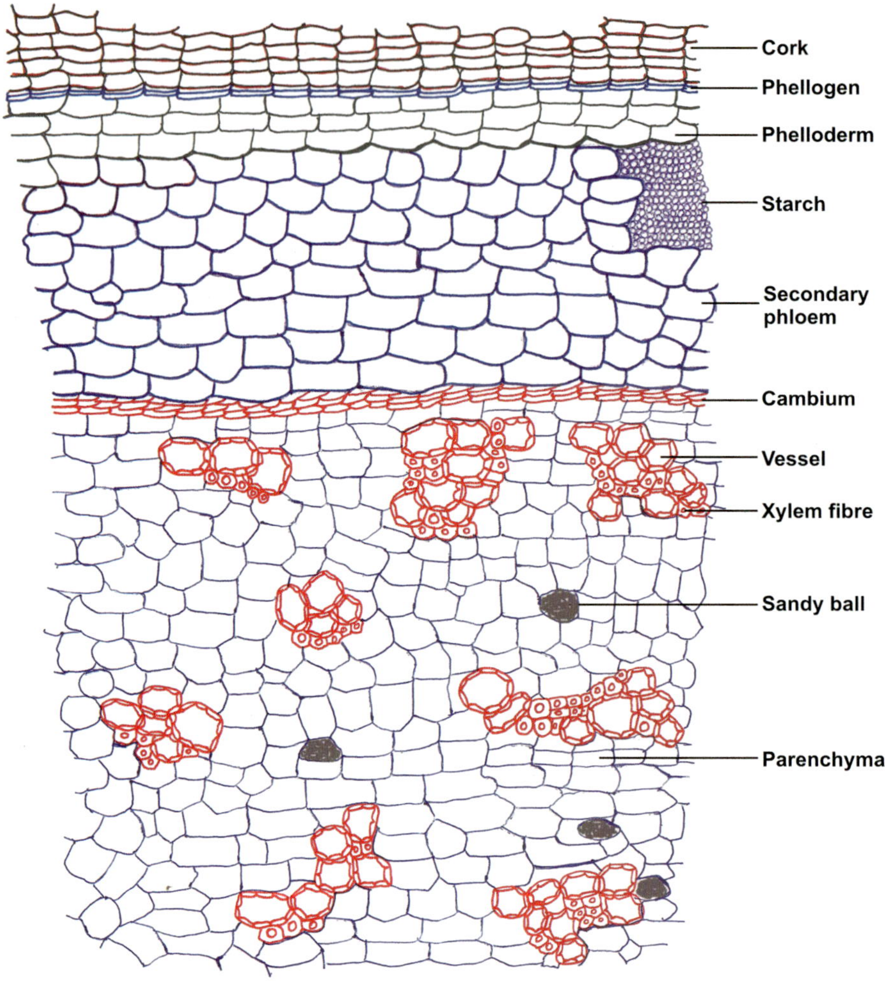

T. S. of Belladonna root

Powder Character:

The powder of belladonna root is pale fawnish brown in colour and possesses a slight odour. It gives a faint, slightly bitter taste. The powder shows the following diagnostic characters microscopically:

a. Starch granules:
These are abundantly seen mainly inside the parenchymatous cells. These show an indistinct or faint, round or slit-shaped hilum. These granules are mostly simple, spherical or occasionally compound with two to four or more components.
b. Parenchyma:
Parenchymatous cells are abundantly seen. These cells mainly come from cork and secondary phloem. These cells are big, ovoid to elongated and thin walled. Some of these cells are filled with microsphenoidal crystals of calcium oxalate. A few crystals are also found as scattered. Some of these cells are filled with starch.
c. Cork cells:
These cells are brown, thick walled, elongated and slightly irregular in surface view.
d. Calcium oxalate:
Numerous sandy balls of calcium oxalate are seen in the cortical cells.
e. Sieve tissues:
Fragments of sieve tissues are seen which are composed of small elongated elements. Some elements show faint sieve areas on the oblique end walls.
f. Xylem vessels and fibres:
These xylem elements are lignified and occur as groups which interlock with each other to form spindle-shaped structures. This is a characteristic feature of belladonna root. The xylem fibres are thin walled and have simple pits. Xylem fibres are usually associated with xylem vessels. Xylem vessels are pitted and occasionally reticulately thickened.
g. Xylem parenchyma:
These parenchymatous cells are lignified and are found scattered as well as associated with the xylem vessels and fibres. Cells are elongated and rectangular, with thick walls and many simple pits.

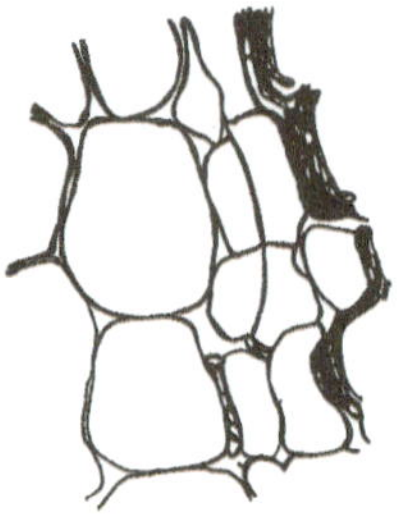
Cork and outer tissues in sectional view

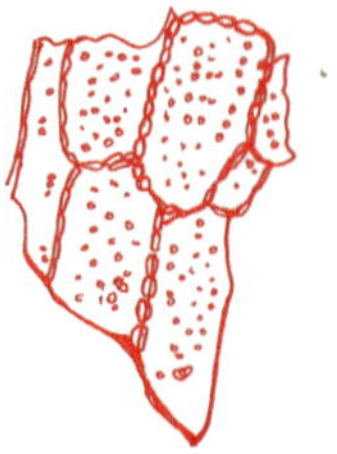
Xylem parenchyma and part of a fibres

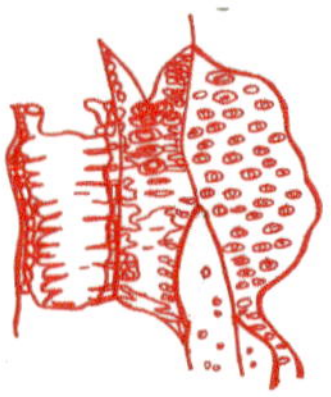
Reticulately thickened and bordered pitted vessels associated with fibre

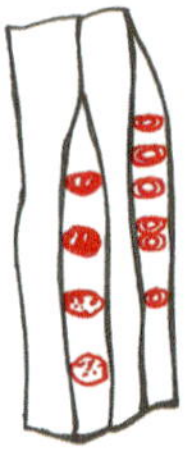
Sieve tissue

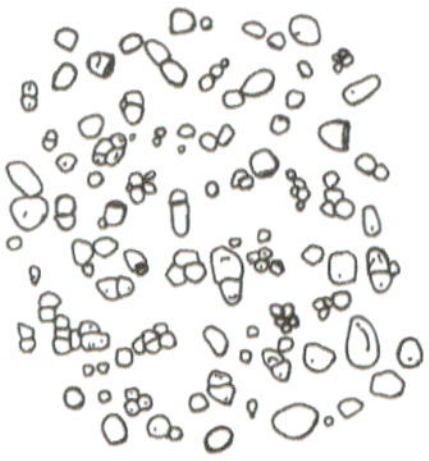
Starch granules

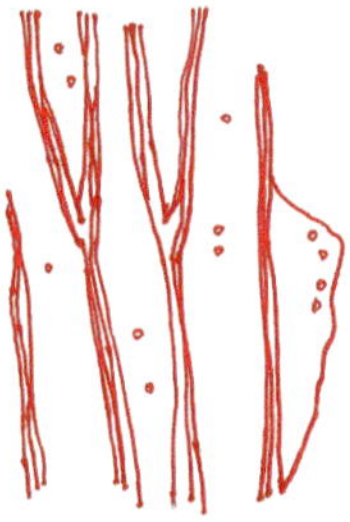
Part of a group of fibres

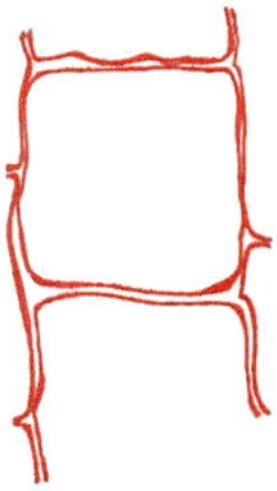
Cells of a medullary ray in R. L. S.

Part of fibre

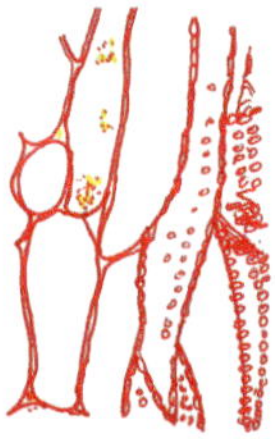
Part of a vessel, fibres and a medullary ray in T. L. S.

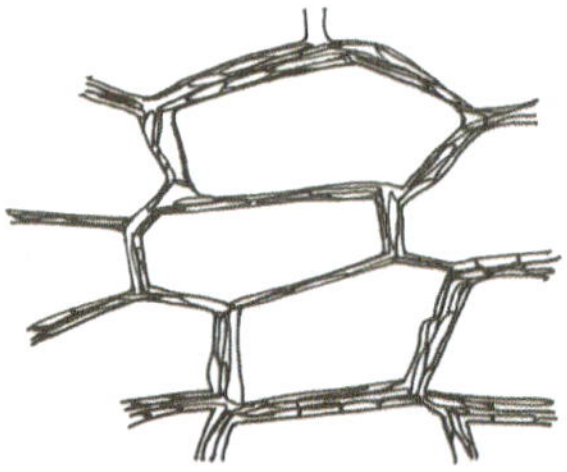
Cork in surface view

Parenchymatous cells containing microsphenoidal crystals of calcium oxalate

9 Cannabis (*Cannabis sativa*)

Synonyms:

Cannabis indica, Indian hemp, Ganja, Marihuana

Biological Source:

It consists of dried flowering tops of the cultivated female plants of *Cannabis sativa* Linn.

Family:

Cannabaceae

Microscopy:

A transverse section of cannabis through a leaflet shows an ordinary dorsiventral structure. It shows the following tissues:

A. Lamina:

   i. Upper epidermis:
   It is composed of cells with straight, anticlinal walls and also possesses numerous trichomes of different types. It bears trichomes which are unicellular, pointed, curved and conical with enlarged bases in which cystoliths of calcium carbonate are seen. Another type of glandular trichomes is observed which are made up of a secreting head of eight radiating club-shaped cells. The oleoresin fluid is secreted by these cells and gets accumulated beneath the cuticle to give it an appearance of a covering envelope on the upper epidermis. Some of the glands have a cylindrical, multicellular and long pedicel and are several celled in diameter, while some glands do not have any stalk. Stomata are of anomocytic type and are numerous.

   ii. Mesophyll:
   It can be differentiated into two separate zones:

      a Palisade:
      It is composed of usually one layer (rarely few layers) of cylindrical cells.

      b. Spongy tissue:
      The spongy parenchyma consists of two to four layers of rounded cells which contain cluster crystals of calcium oxalate.

   iii. Lower epidermis:
   The lower epidermal cells are smaller than those of the upper epidermis and possess conical trichomes which are longer and slender but without cystoliths. It also shows the usual type of glandular hairs especially over the midrib.

B. Midrib:
Transverse section through the midrib shows an arc of xylem and central phloem and the remaining of the tissue is composed of parenchymatous cells. Transversely cut section of the petiole shows a broad collenchymatous hypodermis made up of about eight layers of cells and an arc of vascular bundle at the centre.

Powder Character:

Indian hemp, i.e. cannabis, occurs as green to brown flattened masses which are hard and resinous. The powdered drug is brownish to greenish brown in colour with characteristic, strong odour and slight taste.

The powdered drug shows the following characters:

a. Fragments of bracts:
   Upper epidermis is made up of straight-walled polygonal cells with a slightly striated cuticle. It shows trichomes which are short, conical, unicellular and cystolithic, enlarged at the base where deposits of calcium carbonate are seen. It also bears few small glandular trichomes. Palisade cells are small, closely packed and a few containing cluster crystals of calcium oxalate. The lower epidermal cells are small, with numerous anomocytic stomata and glandular trichomes.
b. Fragments of bracteoles in surface view:
   Upper epidermis is composed of polygonal cells with unevenly thick and beaded walls. The lower epidermal cells are smaller with sinuous, slightly thick and beaded walls. Numerous anomocytic stomata are seen along with conical, unicellular covering trichomes, wide at the base and tapering at the apex. Small cluster crystals of calcium oxalate are seen in the mesophyll.
c. Trichomes:
   These are abundant and scattered. Smaller ones are found attached to pieces of epidermis. Covering trichomes are conical and unicellular, with or without cystoliths. Few are short and enlarged at the base and with pointed apex. Glandular trichomes are numerous and distinctly characteristic. Most of these have a multicellular, multiseriate stalk with a multicellular head containing 8–12 radiating cells. Stalks are cylindrical, three to five cells in diameter. Other less commonly occurring glandular trichomes are smaller and have a uniseriate stalk composed of one or two cells. A spherical head is seen with one to four, occasionally eight, cells.
d. Fragments of the stigmas:
   These are found abundant and orange to reddish brown and have elongated papillae. These papillae are thin walled, cylindrical and rounded at the tip. These get detached and are found scattered in the powder.
e. Calcium oxalate crystals:
   Cluster crystals of calcium oxalate are seen scattered and also within the parenchymatous cells.

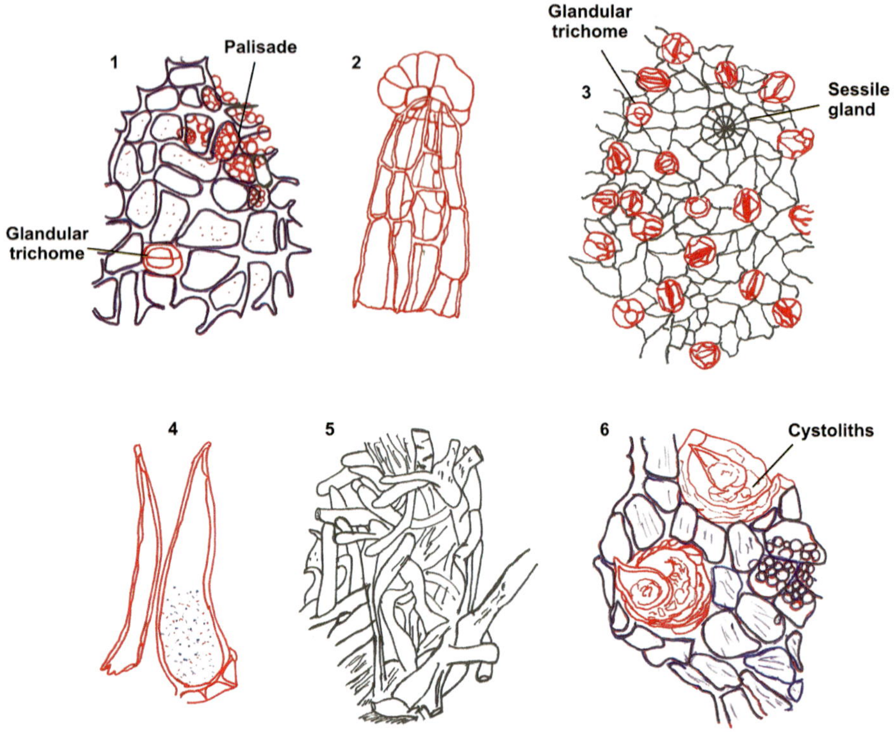

1. Upper epidermis of a bract in surface view showing faint cuticular striations, a small glandular trichome and part of the underlying palisade with some of the cells containing crystals of calcium oxalate
2. A muiticellular, multiseriate glandular trichome
3. Lower epidermis of a bract in surface view showing anomocytic stomata, a small glandular trichome and a sessile gland
4. Covering, non-cystolithic trichomes
5. Part of a stigma showing papillae
6. Upper epidermis of a bract in surface view showing covering trichomes containing cystoliths and part of the underlying palisade

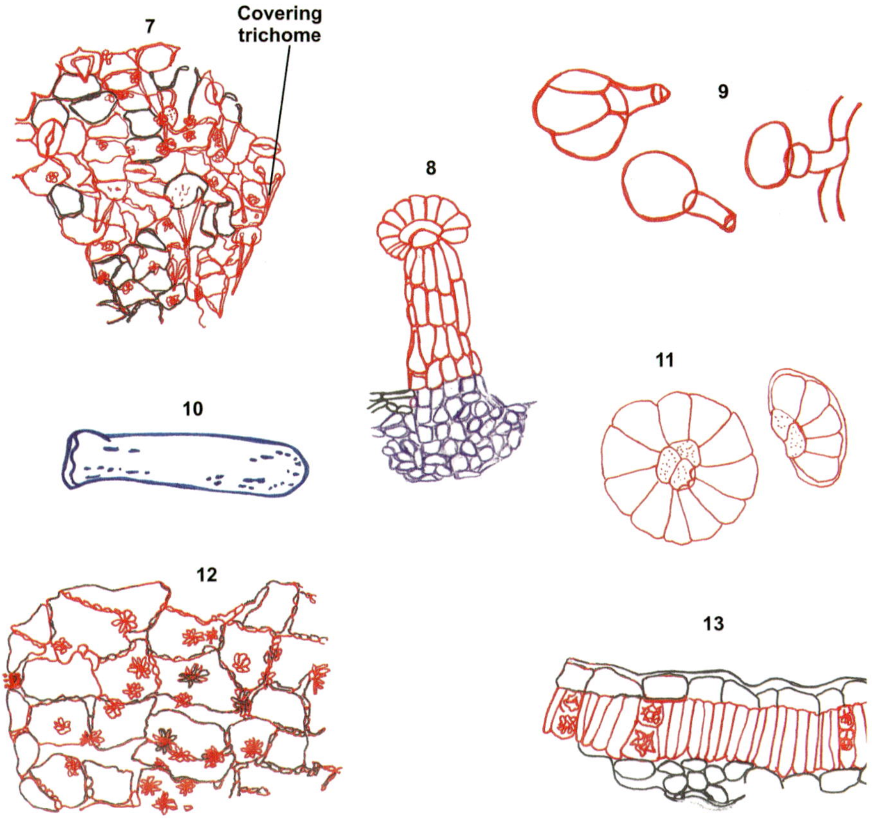

7. **Lower epidermis of a bracteole in surface view showing anomocytic stomata, covering non-cystolithic trichomes and calcium oxalate crystals in the underlying mesophyll**
8. **A muiticellular, multiseriate glandular trichome attached to an epidermis**
9. **Small glandular trichomes**
10. **Part of a papilla from a stigma**
11. **Detached heads from the muiticellular glandular trichomes**
12. **Upper epidermis of a bracteole in surface view showing beaded walls and cluster crystals of calcium oxalate in the underlying mesophyll**
13. **Part of a bract in sectional view showing calcium oxalate crystals in the palisade**

10 Caraway (*Carum carvi*)

Synonyms:

Caraway fruit, Shah jeera, Sushavi

Biological Source:

It consists of dried ripe fruits of the plant *Carum carvi* Linn.

Family:

Umbelliferae

Microscopy:

Caraway shows features of a typical umbelliferous fruit.

Cremocarp:
Cremocarp is a variety of schizocarp (splitting fruit) which divides into two one-seeded portions, each corresponding to one carpel. This carpel itself does not open to liberate the seed; thus, these schizocarps are indehiscent fruits. A cremocarp consists of two parts, each of which is called a 'mericarp'. These two mericarps are connected by a thick-walled sclerenchymatous central stalk called 'carpophore'. A single seed is seen in each mericarp. Raphe is a single ridge of vascular bundle at the middle of the commissural surface. The carpophore is situated just in front of the raphe.

Transverse section of a mericarp shows two prominent surfaces: commissural and dorsal. The commissural surface is flat with two distinct ridges and carpophore in the middle. The dorsal surface shows three ridges. Therefore, each mericarp shows five primary ridges.

A mericarp can be divided into pericarp and seed.

A. Pericarp:
   It is made up of epicarp, mesocarp and endocarp:

   i. Epicarp:
      It is glabrous and composed of thin-walled, polygonal tabular cells with slightly straight anticlinal walls, occasional stomata and striated cuticle.
   ii. Mesocarp:
      It is parenchymatous and surrounds the endosperm. This region shows four vittae on the dorsal surface and two vittae on the commissural side along with five vascular strands in the ridges. Each vascular strand is attached to a pitted sclerenchyma and has a single secretary canal at the outer margin. The six vittae appear somewhat flattened and elliptical in the section. These vittae run from the base of the fruit to the base of the stylopod. These are lined with a layer of small, reddish-brown cells and contain pale yellow oleoresin.
   iii. Endocarp:
      It is made up of a single layer of thin-walled, elongated and sub-rectangular cells. Exact parquetry arrangement of cells is absent, but cells are arranged more or less parallel to one another.

B. Seed:
   It consists of testa, endosperm and embryo:

   i. Testa:
      It is brownish in colour and made up of a single layer of small cells.
   ii. Endosperm:
      The endosperm covers majority of the transverse section and made up of thick-walled, polygonal, cellulosic parenchymatous cells. These thick walls also contain deposits of a β(1, 4)-mannan as a reserve polysaccharide. Endosperm contains oil globules, aleurone grains and very small rosette crystals of calcium oxalate.
   iii. Embryo:
      Endosperm shows presence of embryo when a section is taken through the region near the apex of the mericarp.

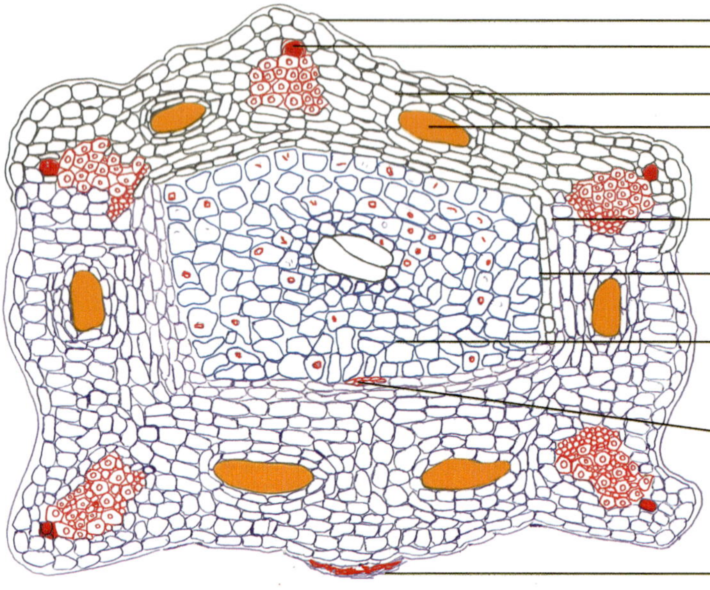

T. S. of mericarp of Caraway fruit

Powder Character:

The powder of caraway fruits is dark brown with characteristic, aromatic odour and slight sweetish taste. Microscopically the powder shows the following characters:

a. Epicarp:
   Fragments of epicarp are seen and composed of a layer of indistinct, colourless polygonal cells with striated cuticle. In surface view, the cells are elongated with thin sinuous walls. It shows many stomata whose long axes appear parallel to those of cells of the epicarp.
b. Vittae:
   Numerous brown fragments of vittae are observed which are composed of polygonal, thin-walled cells. Due to fine cracks, these fragments of vittae appear as broken glass pieces.

c. Endosperm:
It is abundantly seen and contains aleurone grains and micro-rosette crystals of calcium oxalate. Cell walls are sometimes moderately thickened.
d. Sclereids:
Sclereids from the mesocarp occur in large groups. These are placed in a single layer and are associated with thin-walled unlignified parenchymatous cells. Each sclereid is rectangular to sub-rectangular in outline, and walls are moderately or heavily thickened and show regularly placed, distinct, well-marked pits.
e. Testa:
It is found attached to the endocarp and composed of a single layer of brown, thin-walled and polygonal cells.
f. Fragments of endocarp:
These are seen as composed of a layer of large cells with thin, lignified walls. Cells are seen as elongated in surface view with their long axes parallel to each other. These endocarp cell layers are found attached to the pieces of testa.
g. Fibrovascular tissue:
Occasionally small fragments of lignified fibrovascular tissue are observed. These are made up of small, thin-walled fibres and annular and spiral vessels.

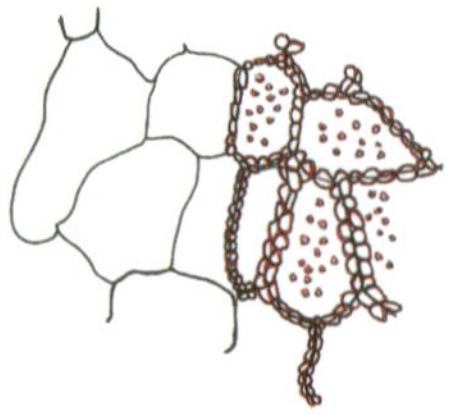

Thicker-walled sclereids with adjacent thin walled parenchyma

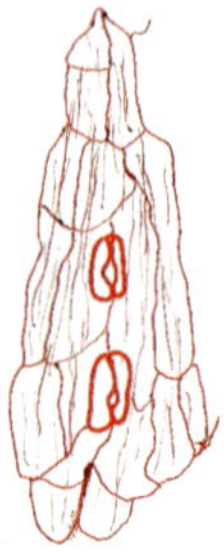

Epicarp in surface view showing stomata and striated cuticle

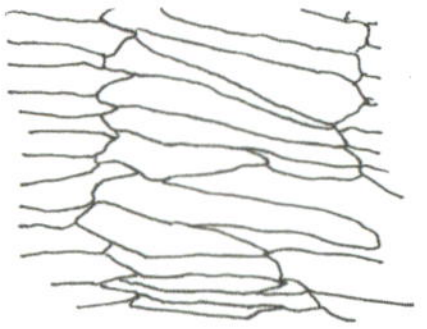

Endocarp in surface view (Parquetry layer)

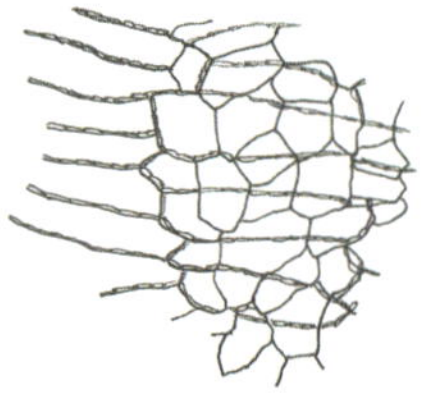

Endocarp in surface view with underlying testa

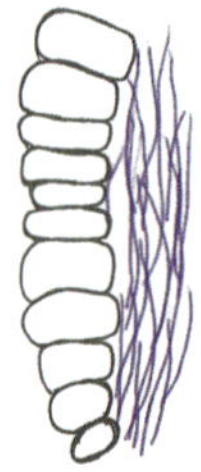

Endocarp and testa in sectional view

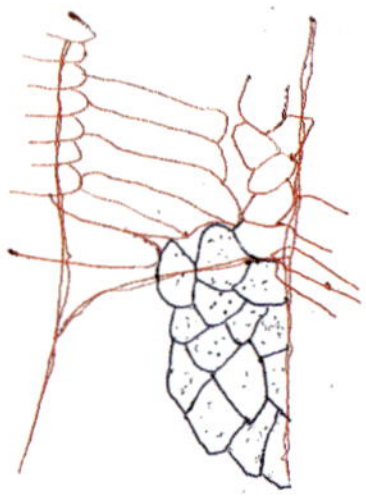

Part of a vitta showing a transverse septum, with underlying endocarp in surface view

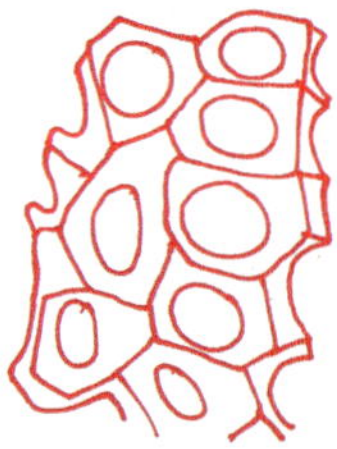

Endosperm containing microrosette crystals of calcium oxalate

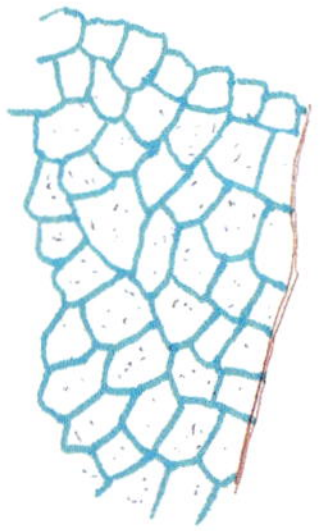

Fragment of vitta

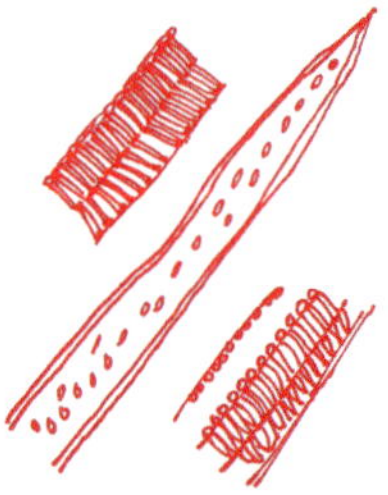

Elements of the fibro-vascular tissue

11 Cinchona (*Cinchona calisaya*)

Synonyms:

Cortex cinchonae, Peruvian bark, Jesuit's bark

Biological Source:

It consists of dried stem bark of the plant *Cinchona calisaya* Wedd., *C. officinalis* Linn., *C. succirubra* Pav. Ex. Klotzsch and *C. ledgeriana* Mocns.

Family:

Rubiaceae

Microscopy:

Transverse section of the cinchona bark shows the following tissues microscopically:

A. Periderm
   It is made up of cork, phellogen and phelloderm:

   a. Cork:
      It consists of many layers of thin-walled cells arranged in regular radial rows. Cells appear as flat and polygonal with reddish-brown cell contents. The cell walls are suberised.
   b. Phellogen:
      It is made up of two to three layers of thin-walled rectangular cells.
   c. Phelloderm:
      It is placed within the cork cambium. It is made up of several (up to eight) layers of regular, thin-walled rectangular cells with dark walls and without any cell contents. Cork cambium is not distinctly seen in commercial samples.

B. Cortex:
   This portion of the bark is wide and consists of many layers of tangentially elongated, thin-walled cells. These are made up of cellular parenchyma and walls are reddish brown. Some of the cells of the cortex are filled with microsphenoidal crystals of calcium oxalate. Sometimes idioblasts, containing microcrystals (mostly prisms) of calcium oxalate and secretary cells (cavities or secretion canals or latex ducts), are also found as scattered in the layers of the cortex. A few cells show presence of minute starch granules contained within them. These are mainly observed near the inner border of the cortical parenchyma. Sometimes these are large enough and spaced at specific intervals. These appear oval in the transverse section.
C. Secondary phloem:
   This region is made up of sieve tubes, phloem parenchyma, phloem fibres and medullary rays:

   i. Sieve tubes:
      The end walls of the sieve tubes are seen at the right angles to the axis as the component cells are long and wide. The companion cells are narrow. Sieve tubes are collapsed and compressed most of the time in commercial samples of the bark.
   ii. Phloem parenchyma:
      It resembles the cortical parenchyma in many respects. It consists of dark reddish-brown thin walls. A few of these cells show presence of microprisms of calcium oxalate within them.
   iii. Phloem fibres:
      These are many, large, fusiform, thick walled and lignified. These phloem fibres occur as single or in irregular radial rows, as groups of two to five

fibres. These are seen as intermingled with phloem parenchyma and in between the medullary rays. Distribution and size of the phloem fibres differ in different species of cinchona bark and thus can indicate the specific striking, characteristic features of a particular species and can help in identification. Many times these groups of fibres occur as rounded, oval or spindle shaped. The thick walls of fibres are striated and show conspicuous tubular or funnel-shaped pits. These fibres appear yellowish in colour and have a small lumen.

iv. Medullary rays:
These rays run radially transversing the phloem parenchyma. These are one, two or three seriate. These are narrow, thin walled and almost straight and run up to the cortex. The cells are thin walled and somewhat radially elongated. Some of these cells of the medullary rays contain starch grains.

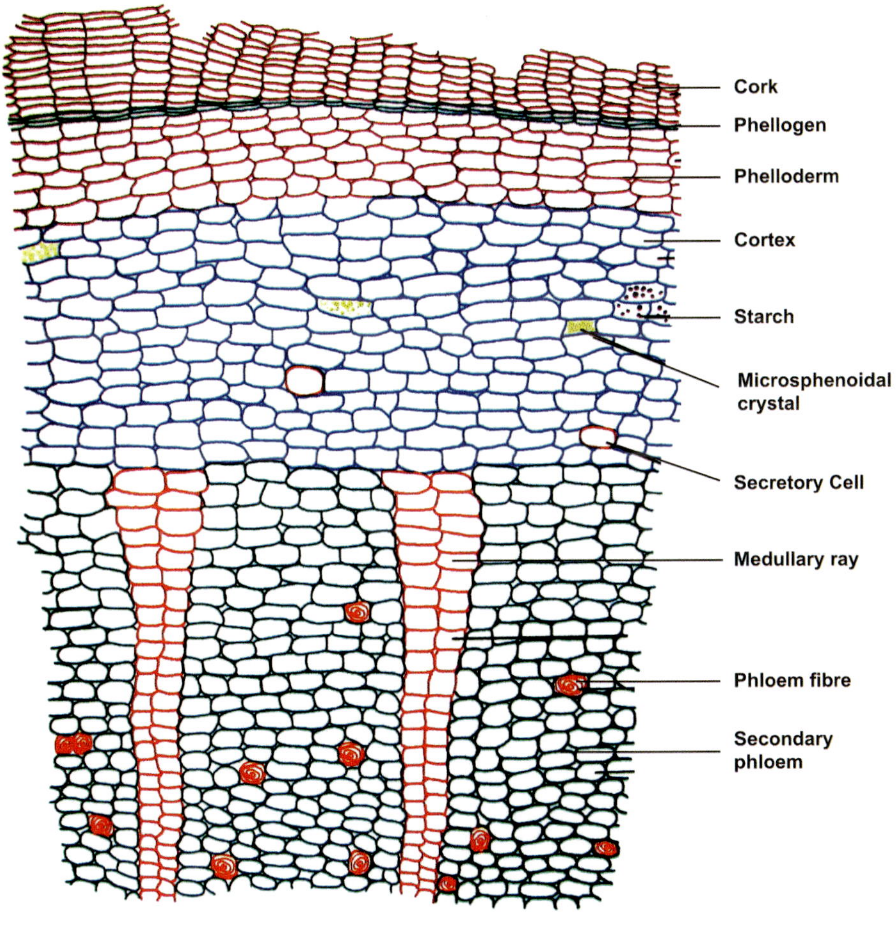

T. S. of Cinchona bark

Powder Character:

The powder of cinchona bark is reddish brown with a slight, characteristic odour and bitter, astringent taste.

The powder microscopically shows the following characters:

a. Fibres:
These phloem fibres are many, yellowish, fusiform, large, lignified and fragmented. These occur as isolated or in groups of two or three. Individual fibres have bluntly pointed ends. The walls are thick and show striations. Walls are strongly lignified and possess simple or branched pores. The lumen is small, uneven and short. The walls have numerous pits which are distinct and funnel shaped and which open into the small lumen of the fibre. Sometimes longitudinal tissues are observed in the walls at intervals.
b. Cork:
Cork cells are seen as thin walled, flat, polygonal and suberised. These cells contain reddish-brown matter within them. Generally numerous fragments of cork cells are observed.
c. Parenchyma:
These parenchymatous cells are abundant and arise from phloem parenchyma and medullary rays. These cells appear yellowish to reddish brown in colour. Phloem parenchyma cells are thin walled and fragmented. Few of these cells show presence of some colouring matter along with small starch granules, and some contain microprisms of calcium oxalate. The parenchymatous cells of medullary rays are mainly associated with fibres and possess slightly thick walls.
d. Calcium oxalate crystals:
These are mainly observed in some of the parenchymatous cells. Isolated or scattered crystals are very small and irregular in shape. Generally microprisms of calcium oxalate are observed in the powder.
e. Starch grains:
These are within some of parenchymatous cells and some as scattered. These are small, simple and spherical or rarely compound with two to five components.
f. Stone cells and cluster crystals of calcium oxalate are absent.

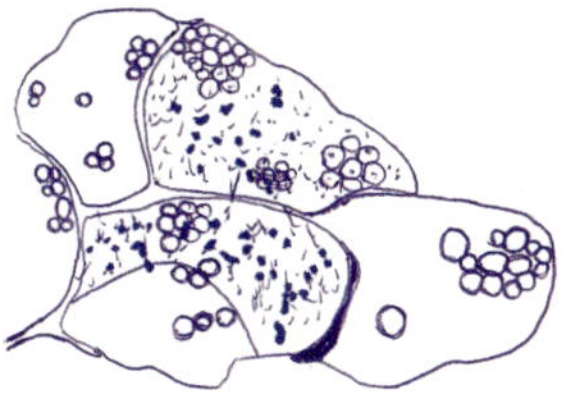

Parenchymatous cells containing starch granules and brown pigments

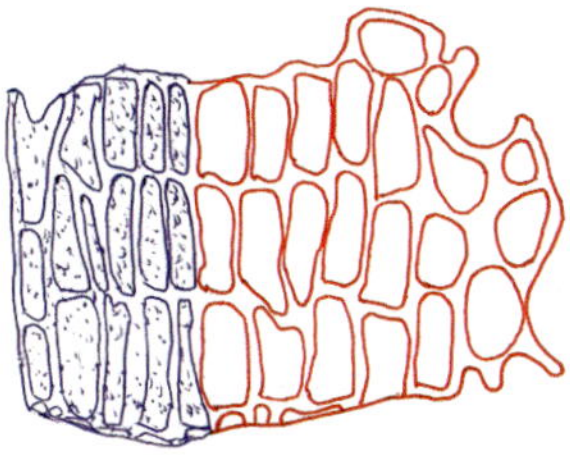

Cork and phelloderm in sectional view

Single fibre

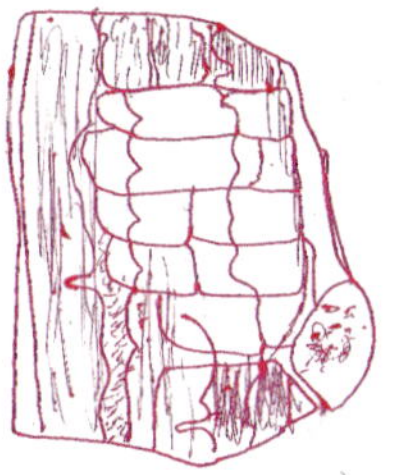

Part of a group of fibres and phloem parenchyma with overlying medullary rays in R. L. S.

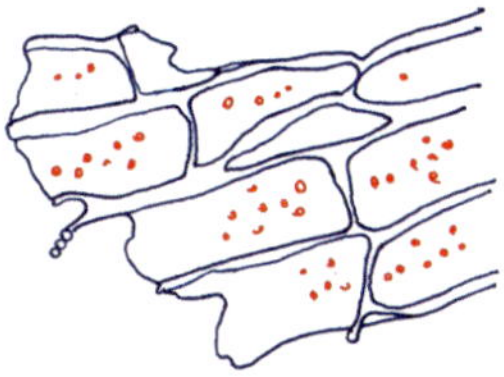

Phloem parenchyma with pits

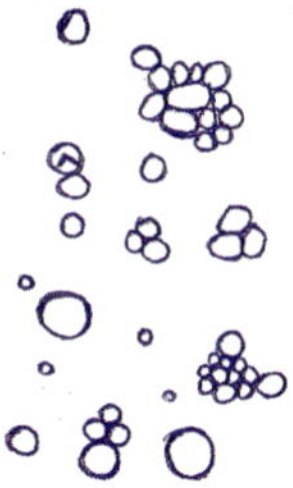

Starch granules

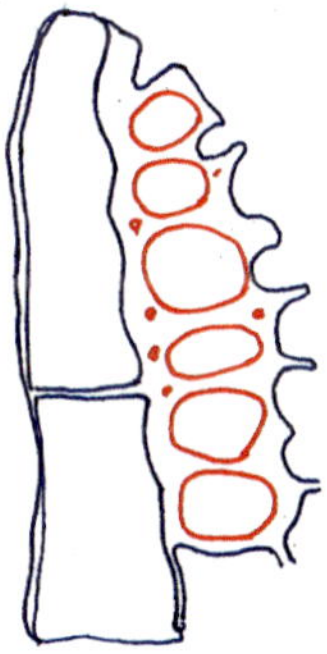

Phloem parenchyma and part of medullary ray in T. L. S.

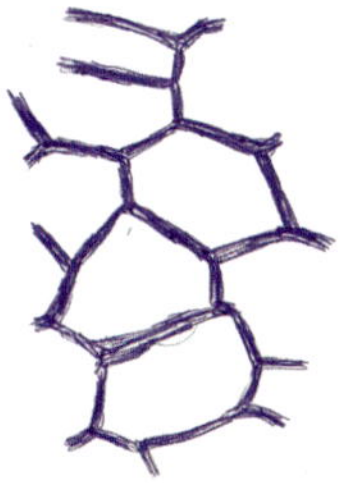

Cork in surface view

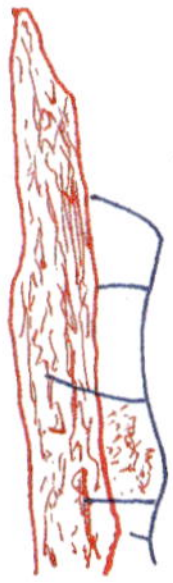

Part of fibre with phloem parenchyma, one cell containing calcium oxalate microcrystals

12 Cinnamon (*Cinnamomum zeylanicum*)

Synonyms:

Ceylon cinnamon, Cinnamon bark, Dalchini, Twak

Biological Source:

It consists of dried inner bark of the shoots of coppiced trees of *Cinnamomum zeylanicum* Nees.

Family:

Lauraceae

Microscopy:

Microscopically transverse section of the cinnamon bark shows the following tissues:

A. Periderm:
   Few layers of cork cells can be seen in which the outer cells are thin walled and inner cells are lignified and thick walled. Phellogen and phelloderm are not distinguishable from each other or from cork.
B. Cortex:
   This is many times absent in commercial samples of the bark. If present it is found as patches and composed of 10–15 layers of parenchyma with scattered sclereids. Each sclereid is rectangular and pitted. Some of the parenchymatous cells contain minute acicular raphides and abundant starch.

C. Pericycle (stone cell layer or sclerenchymatous band):
In most of the samples, the outermost limit of the bark is marked by a pericycle which produces the light coloured, wavy, longitudinal lines on the outer layer of the bark. The pericycle is composed of a continuous ring of three to four layers of sclereids with small groups of pericyclic fibres embedded in it at intervals:

   i. Sclereids:
   These are pitted, lignified, thick walled and isodiametric, with well-defined pit canals. Sclereids are many times more thickened upon the inner walls than upon the other three, giving them a characteristic 'U-shaped' appearance. They contain a few starch grains.
   ii. Pericyclic fibres:
   Small groups of about 6–15 long, lignified, pericyclic fibres occur at intervals. These fibres have strongly thickened lignified walls with stratification and pit canals.

D. Secondary phloem:
It consists of phloem parenchyma, fibres and medullary rays:

   i. Phloem parenchyma:
   It consists of sub-rectangular, thin-walled cells with starch grains (both simple and compound) and numerous acicular crystals of calcium oxalate. Some of the phloem parenchyma cells contain tannin. Oil cells and mucilage cells are also observed in phloem parenchyma. Idioblasts are somewhat longitudinally elongated and contain volatile oil or mucilage. The sieve tube tissue which is embedded in the phloem parenchyma is often obliterated.
   ii. Phloem fibres:
   The phloem fibres occur singly or in short tangential rows of 2–5 and are more abundant towards the inner part of the bark. These are circular and slender and their thick lignified walls show stratifications. The width of phloem fibres and size of starch grains are important identifying characters of cinnamon, especially a distinction from cassia bark.
   iii. Medullary rays:
   The secondary phloem is divided up by radial medullary rays. These are uni- or biseriate near the cambium but become broader towards the outer layer by tangential growth of cells. The rays are 7–14 cells high. The medullary ray cells are radially elongated and thin walled with yellow–brown cell contents containing numerous acicular crystals of calcium oxalate.

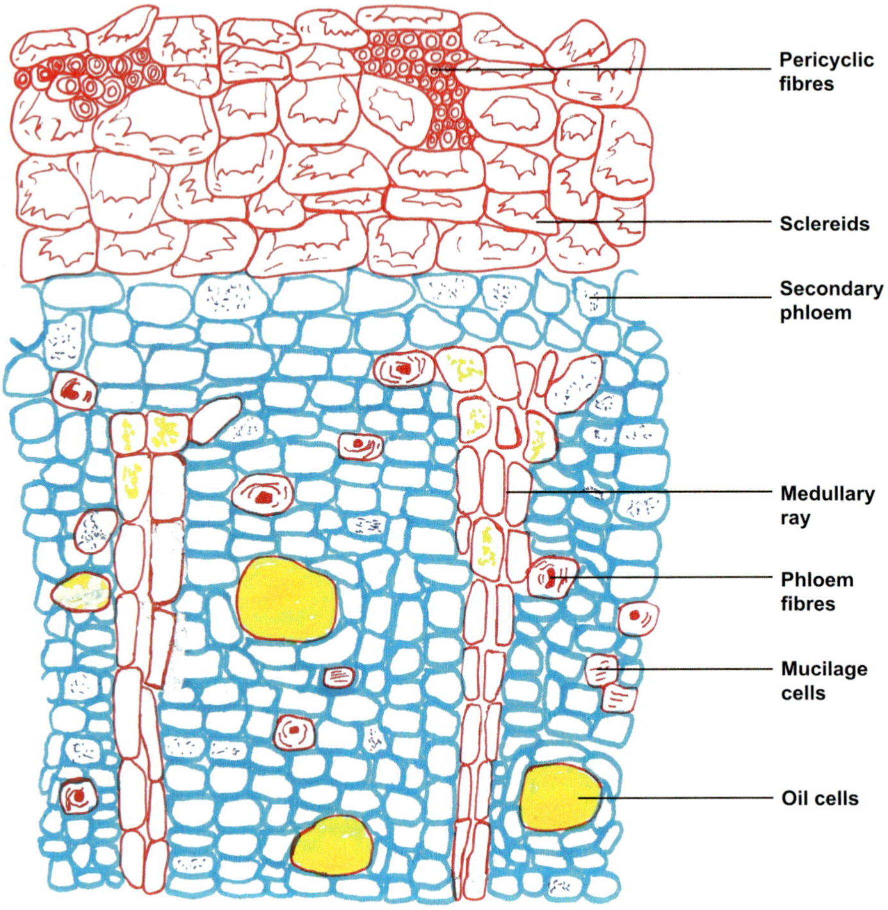

T. S. of Cinnamon bark

Powder Character:

The powder of the cinnamon bark is reddish brown in colour with a characteristic, pleasant, sweet fragrance and taste. It shows the following diagnostic characters microscopically:

a. Fibres:
   These are abundant and usually occur singly. These have thick, lignified, stratified walls and small, somewhat uneven, narrow lumen and few inconspicuous slit-shaped pits. Occasionally fibres are found along with the sclereids of the pericycle, and a few occur associated with the oil cells and parenchyma of the phloem.
b. Sclereids:
   These are abundantly seen which occur singly or in small groups. These are of various sizes and shapes but usually are isodiametric, with thick walls, lignified; the outer wall is thinner than the other walls. Most of the cells are thick walled which gives a characteristic U shape; the lumen is small. Pits are numerous and conspicuous; striations are usually visible.
c. Starch grains:
   These are numerous, found scattered and inside parenchymatous tissues or sclereids. These are commonly observed in phloem parenchyma and medullary rays. These are small, single or compound with four or more components. A rounded or slit-shaped hilum is observed in a few large grains. Diameter of the grains is less than 10 μ, which is a distinguishing character from the cassia bark.
d. Calcium oxalate crystals:
   The thin-walled phloem parenchyma and medullary ray cells of the phloem show small, numerous acicular crystals of calcium oxalate.
e. Oil cells:
   Oil cells are seen as entire or as fragments. These cells are often associated with the parenchyma or fibres of the phloem; cells are large and ovoid and usually occur singly.
f. Cork:
   The cork cells are usually absent. Very occasionally fragments of cork can be seen. These cells are thin walled and polygonal in surface view. In sectional view, fragments show the cell layers arranged in alternating bands of thin-walled cells and thick-walled, indistinct lignified cells.

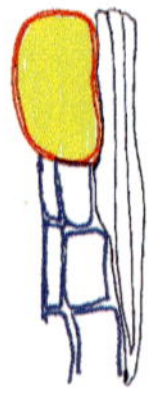

Part of a fibre with an associated oil cell and phloem parenchyma

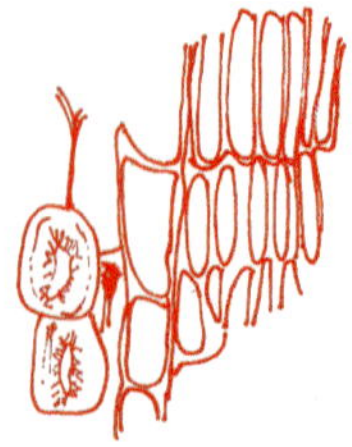

Part of the cork and cortex in sectional view

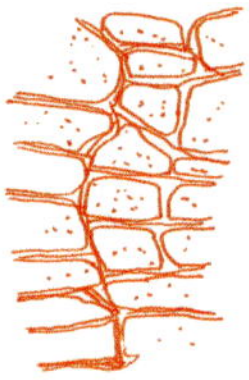

Cork in surface view

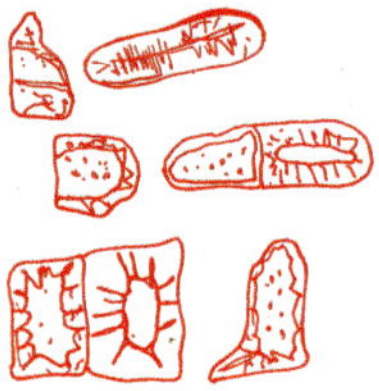

Sclereids

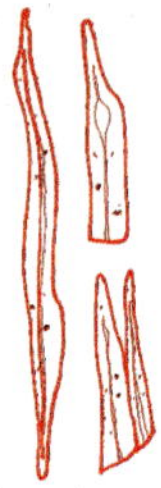

Fibres and part of small group of fibres

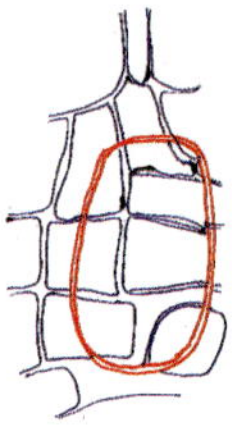

Phloem parenchyma and an oil cell

Calcium oxalate crystals

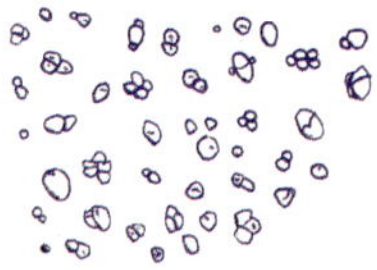

Starch granules

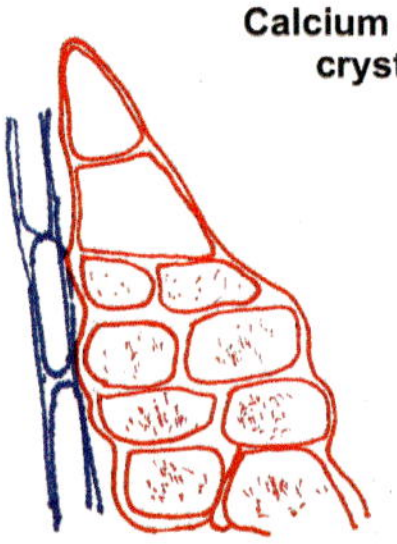

Part of medullary ray with some of the cells containing acicular crystals of calcium oxalate and associated phloem parenchyma in T. L. S.

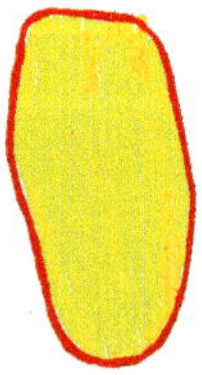

A single oil cell

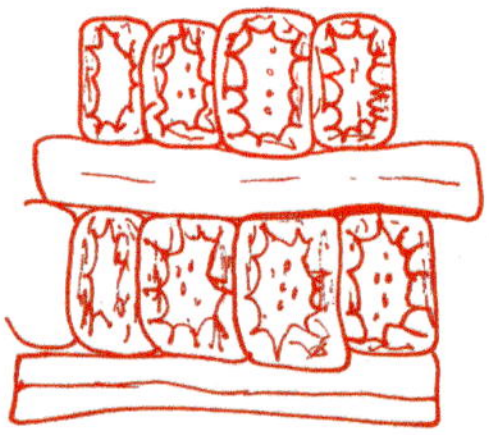

Part of a group of fibres and sclereids from the pericycle

13 Clove (*Eugenia caryophyllus*)

Synonyms:

Caryophyllum, Clove flower, Clove buds, Laung, Lavang

Biological Source:

It consists of dried flower buds of the plant *Eugenia caryophyllus* Sprange.

Family:

Myrtaceae

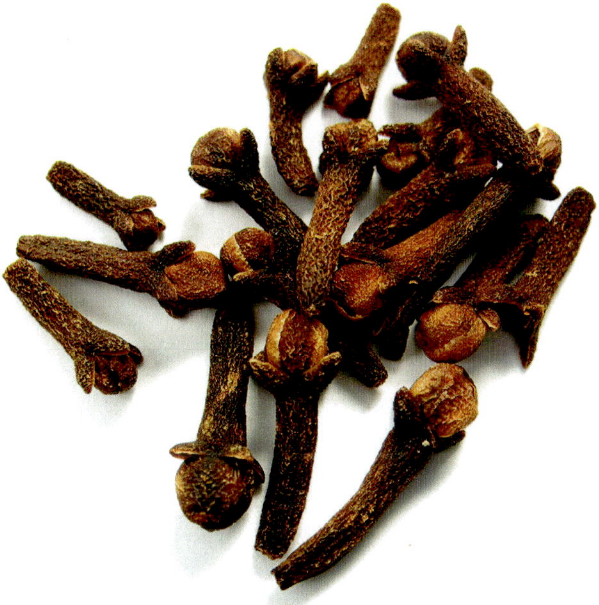

Microscopy:

In the case of clove bud, a transverse section is taken through the ovary as well as through the hypanthium, i.e. stalks of the bud. The short upper portion present immediately below the calyx contains bilocular ovary. The lower portion, i.e. stalk of the bud (hypanthium) that lies at the lower part, is long, solid and sub-cylindrical. The transverse section through the hypanthium shows the following tissues microscopically from the periphery towards the centre:

A. Epidermis:
   It is made up of a single layer of small, tabular cells with straight walls and highly cuticularised. It shows presence of anomocytic (ranunculaceous) type of

stomata. These stomata appear as slightly raised above the epidermal surface. The substomatal spaces are prominent and well defined.

B. Cortex:
It occupies the major portion and can be divided into three different zones:

   i. Outer (upper, peripheral) zone:
   It shows two to three layers of big, ellipsoidal, schizolysigenous oil glands embedded in radially elongated parenchymatous cells. Oil glands have long radial axis and an epithelium which is composed of two or three layers of flattened cells. Parenchymatous cells contain tannins and thus show dark colouration with ferric chloride solution (alcoholic). This staining is also observed with alcoholic osmic acid. Many of the parenchymatous cells show cluster crystals of calcium oxalate.
   ii. Middle zone:
   Within the oil gland layer, a zone of thick-walled cells is seen. Within these cells, a ring of bicollateral vascular bundles is embedded. About 20–25 bundles are present in the ring. The ground tissue contains cluster crystals of calcium oxalate. The vascular bundles are enclosed in an incomplete ring of lignified pericyclic fibres. The xylem is composed of three to five lignified spiral vessels.
   iii. Inner (lower) zone:
   It is made up of loosely arranged parenchymatous cells (aerenchyma) composed of air spaces. Air spaces are separated by lamellae which are thin and one cell thick. This region supports the central columella.

C. Columella:
It forms the central cylinder which is parenchymatous and contains calcium oxalate crystals. A ring of about 15–20 small vascular bundles is seen towards the periphery.

The transverse section through the ovary shows all tissues described above. But instead of the central columella, a bilocular ovary is present at the centre. The ovules are numerous, separated by an axile placentation. The dissepiment of the ovary is parenchymatous. The placentae show cluster crystals of calcium oxalate and vascular bundles.

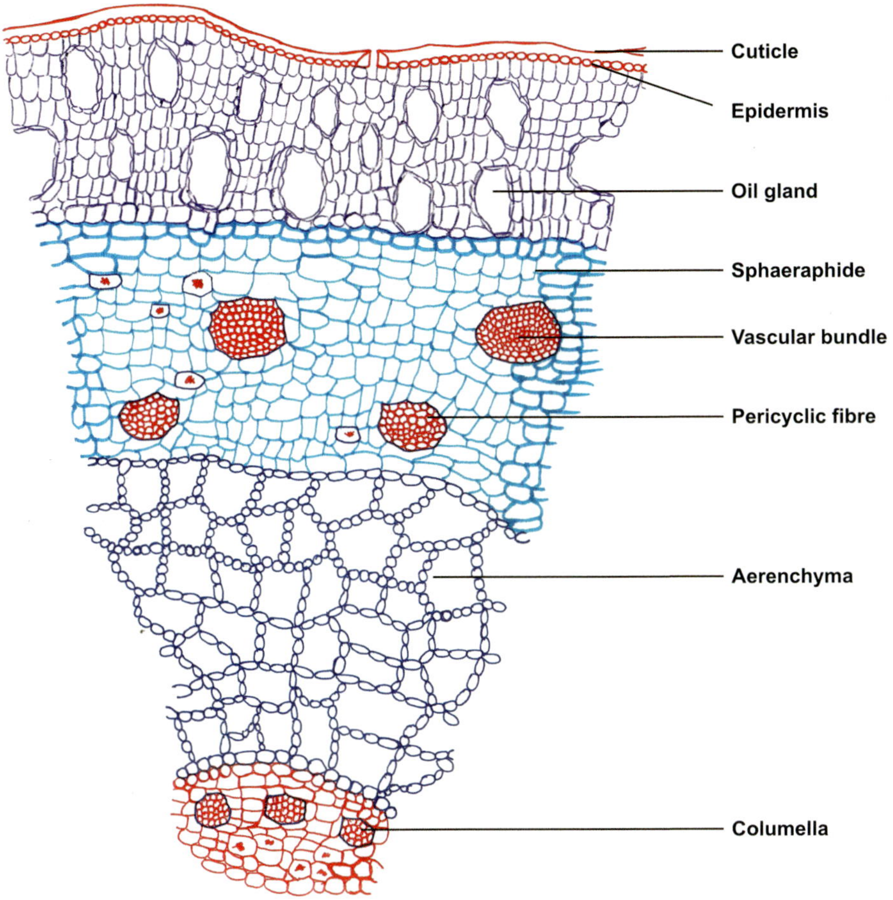

T. S. of Clove flower bud

Powder Character:

The powder of the clove buds is dark brown in colour and possesses a characteristic, spicy aroma and pungent, slightly, characteristic, aromatic taste. The powder shows the following characters microscopically:

a. Hypanthium:
   The fragments of hypanthium and the epidermis occur abundantly. The epidermis is made up of small, polygonal, thick-walled cells. It shows circular, big, anomocytic stomata along with large, brown, ovoid oil glands. A few cluster crystals of calcium oxalate are also observed. The fragments of hypanthium show a thick cuticle.
b. Oil glands:
   These are numerous in the hypanthium, ellipsoidal to ovoid, large, brown and schizolysigenous associated with other parenchymatous tissues.
c. Parenchyma:
   The yellowish-brown parenchyma of hypanthium is abundant. Oil glands are found as embedded in this area. Cells are thickened and sometimes are collenchymatous. It shows a few small cluster crystals of calcium oxalate.
d. Calcium oxalate crystals:
   The cluster crystals of calcium oxalate (sphaeraphides) are found in the parenchymatous cells. These crystals are of various sizes, rarely found scattered and made up of many small components.
e. Fibres:
   These are sclerenchymatous, found singly or in groups of two to three in each group. These are short, broad and bluntly pointed. These show lignified thick walls with faint striations and small pits. The lumen is sometimes filled with brownish matter. These fibres are generally associated with parenchymatous cells or with small groups of vessels (fibrovascular bundle).
f. Aerenchyma:
   Fragments of aerenchyma (loosely packed parenchyma) of the hypanthium are occasionally observed. These are made up of chains of two or three thick-walled parenchymatous cells. These chains show small intercellular air spaces separated by lamellae.
g. Sclereids:
   These are from stalk and appear as oval to sub-rectangular with thick, striated walls and simple or branched pits. Some brownish matter is found inside the lumen.
h. Pollen grains:
   These are small and biconvex with rounded or triangular shape and smooth exine. The immature pollen grains are found inside the pollen sacs.
i. Starch grains:
   Starch grains are not observed in the case of the powder of clove buds. If these are observed in the powder, these come from mother cloves.

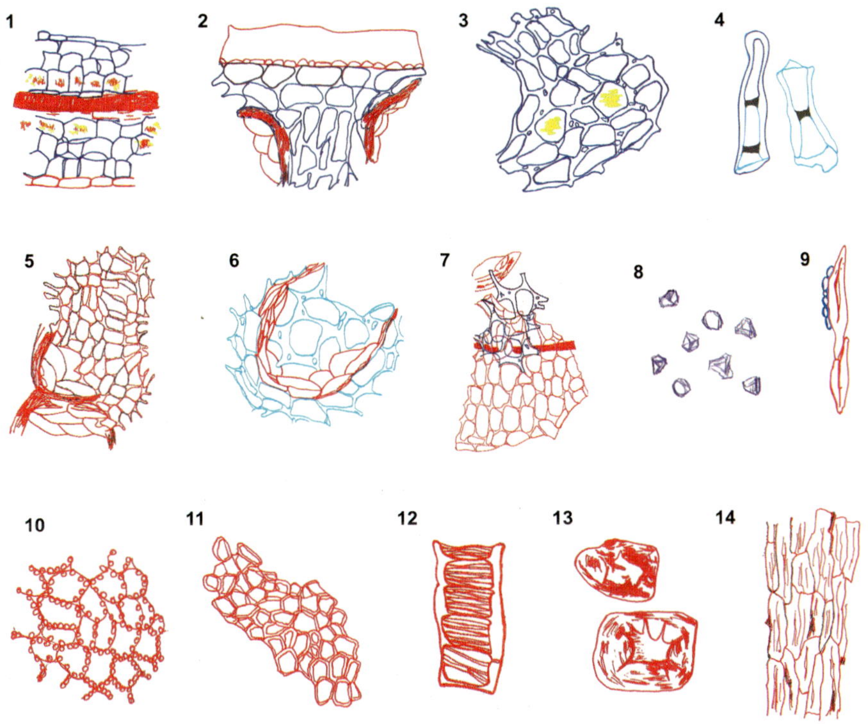

1. Part of the filament of the anther in longitudinal section, showing the central vascular strand and parenchymatous cells containing cluster crystals of calcium oxalate
2. Part of the hypanthium in sectional view showing the thick cuticle, epidermis and underlying parenchyma with oil glands
3. Parenchyma of the hypanthium with cluster crystals of calcium oxalate
4. Fragments of aerenchyma from the hypanthium showing pits
5. Epidermis of the hypanthium in surface view showing a stoma and underlying oil glands
6. Parenchyma of the hypanthium showing an oil gland
7. Epidermis of a petal in surface view with underlying tissue composed of parenchymatous cells containing cluster crystals of calcium oxalate, part of an oil gland and part of a vascular strand
8. Mature Pollen grains
9. Fibrous layer of the anther in surface view
10. Part of group of immature pollen grains
11. Fibrous layer of the anther in sectional view
12. Sclereids from the stalk
13. Epidermis of the filament of the anther in surface view
14. Fibrous and associated parenchymatous cells

14 Colchicum (*Colchicum autumnale*)

Synonyms:

Colchicum, Meadow saffron, Autumn crocus

Biological Source:

It consists of colchicum seeds and corms derived from the plant *Colchicum autumnale* Linn.

Family:

Liliaceae

Microscopy:

Powdered seeds of colchicum appear brownish in colour, are odourless and have a bitter and unpleasant taste. It shows the following tissues microscopically:

A. Starch granules:
   These are abundantly observed as small and mostly simple. Occasionally compound granules with two components also occur. Each granule is spherical to polyhedral with distinct cleft or radiate hilum.

B. Testa:
It is composed of layers of parenchymatous cells with brownish walls. The cells of the outer region are rectangular to polygonal, big and thick walled. The cells of the middle layer are smaller and rounded with unevenly thickened walls and show conspicuous beading. Cells also show characteristic, rounded intercellular spaces within. The cells of the inner region are thin walled, small, regularly placed and rectangular to polygonal in shape. These are compactly packed with no intercellular spaces within. These three layers are seen as attached to other layers and the inner layer is also associated with pigment layer cells.

C. Pigment layer:
Usually fragments of the pigment layer are seen. This layer is composed of a single layer of thin-walled cells. These cells appear rectangular in surface view. These are filled with dark brown pigment, but the cell walls are colourless. This pigment layer is usually seen as associated with outer endosperm or inner parenchymatous layer of testa.

D. Endosperm:
Numerous fragments of endosperm are observed. Cells are large and rectangular with characteristic pitted (porous) walls. These large pits appear circular or oval in surface view. This pitting is less frequent in the cells of the outer layers. Cells also show aleurone grains and fixed oil globules.

E. Strophiole:
It is composed of thin-walled parenchymatous cells filled with starch granules. These cells appear as rounded to rectangular with irregular intercellular spaces.

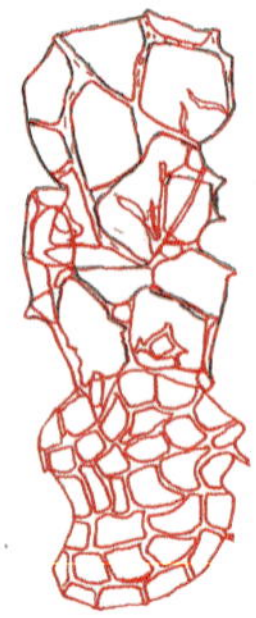

Outer, middle and inner parenchymatous layers of the testa in surface view

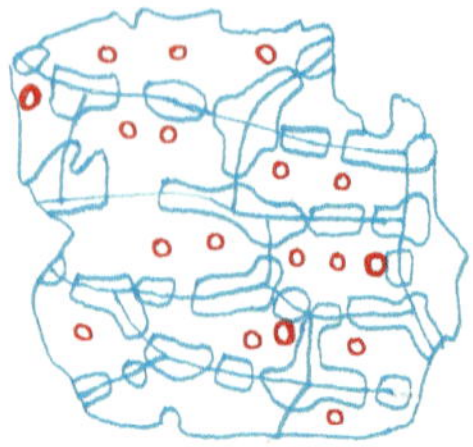

Part of the endosperm showing pits

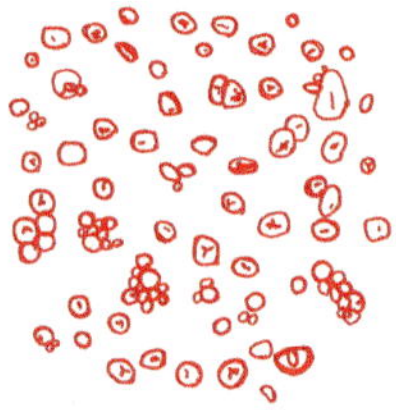

Starch granules

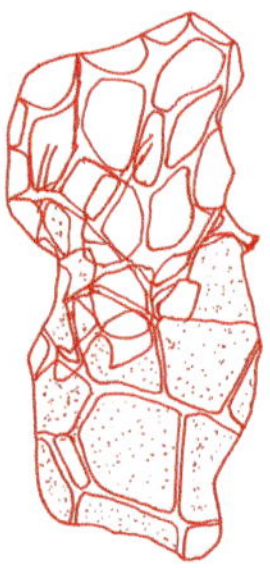

Inner parenchymatous layer of the testa with underlying pigment layer in surface view

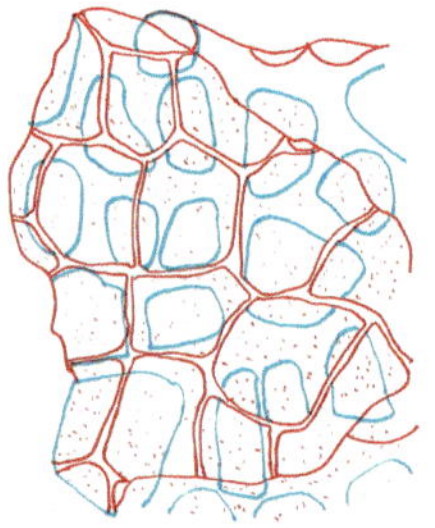

Pigment layer in surface view with underlying endosperm cells

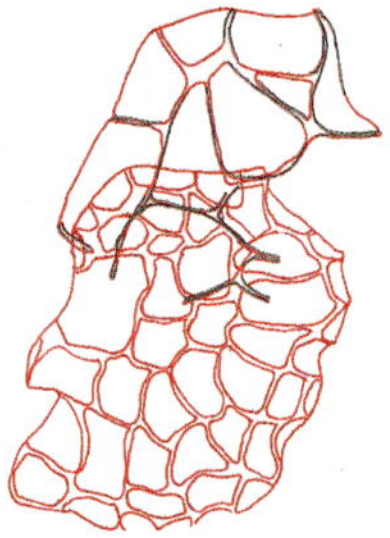

Middle layer of the testa and underlying inner parenchymatous layer in surface view

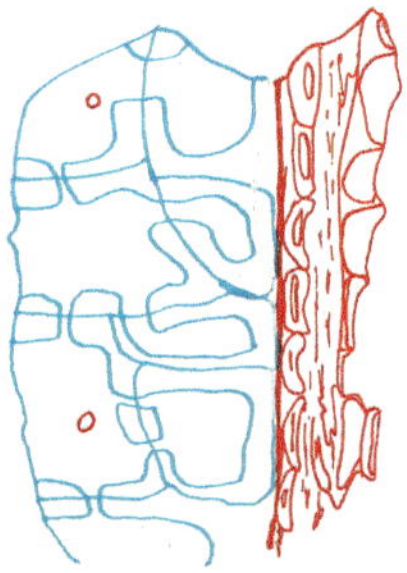

Part of the parenchyma, pigment layer and endosperm in sectional view

Part of a group of vessels from the raphe

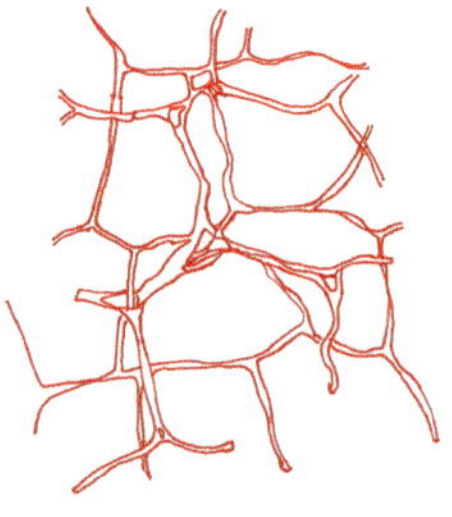

Thin-walled parenchyma of the strophiole

15 Digitalis (*Digitalis purpurea*)

Synonyms:

Folia digitalis, Foxglove leaves

Biological Source:

It consists of dried leaves of the plant *Digitalis purpurea* Linn.

Family:

Scrophulariaceae

Microscopy:

Microscopically a transverse section of a foxglove (*Digitalis purpurea*) leaf shows a typical dorsiventral pattern and the midrib strongly convex on the lower portion of the leaf.

The important tissues in the lamina and midrib region are as follows:

A. Lamina:

   a. Upper epidermis:
   It is single layered with polygonal, straight to slightly wavy-walled cells having smooth and distinct cuticle. It bears both covering and glandular trichomes abundantly. A few stomata of anomocytic type are also observed. Covering trichomes are uniseriate, composed of three to five cells, with acute apex or sometimes bluntly pointed and finely wavy cuticle. Sometimes trichomes

have a few collapsed cells alternating at right angles, and on each marginal tooth of cells, there is usually one or rarely two water pores seen. Glandular trichomes are mainly gathered over the veins and possess a short unicellular or occasionally a short uniseriate stalk and a bicellular or rarely unicellular head; sometimes a uniseriate stalk of 3–4 cells and a unicellular spherical head are seen. A solution of Sudan red in glycerin can be used for staining cuticle of trichomes and epidermal cells as red.

b. Mesophyll:
   Mesophyll can be differentiated into palisade and spongy parenchyma. Palisade is composed of a single layer of compact, radially elongated cells below the upper epidermis. Spongy parenchyma is made up of several (4–6) layers of loosely packed cells and shows many distinct obliquely cut veinlets in this area.
c. Lower epidermis:
   It is similar to the upper epidermis. Stomata and trichomes are more in number than the upper epidermis. The cells of the lower epidermis are polygonal with smooth cuticle but with strongly wavy anticlinal walls.

B. Midrib:
   A transverse section through the midrib region shows that it is more convex below and the epidermal layers of lamina run over the midrib also. Below the upper epidermis and above the lower epidermis, thin collenchymatous layers are seen and the remaining tissue is filled with thin-walled cortical parenchymatous cells. At the central region, an arc-shaped vascular bundle is present, more towards the ventral surface of the midrib. The vascular bundle is encircled by an endodermis containing abundant starch granules. Within this endodermis, a band of collenchymatous pericycle is observed. A narrow phloem is clearly seen on the dorsal surface, and a well-developed, shallow, gutter-shaped radiate xylem tissue is observed towards the ventral surface of the midrib along with some parenchyma cells on the upper side.

   Surface preparations of the leaf show anomocytic stomata, less wavy upper epidermal walls and more strongly wavy walls of lower epidermis along with numerous covering and glandular trichomes.

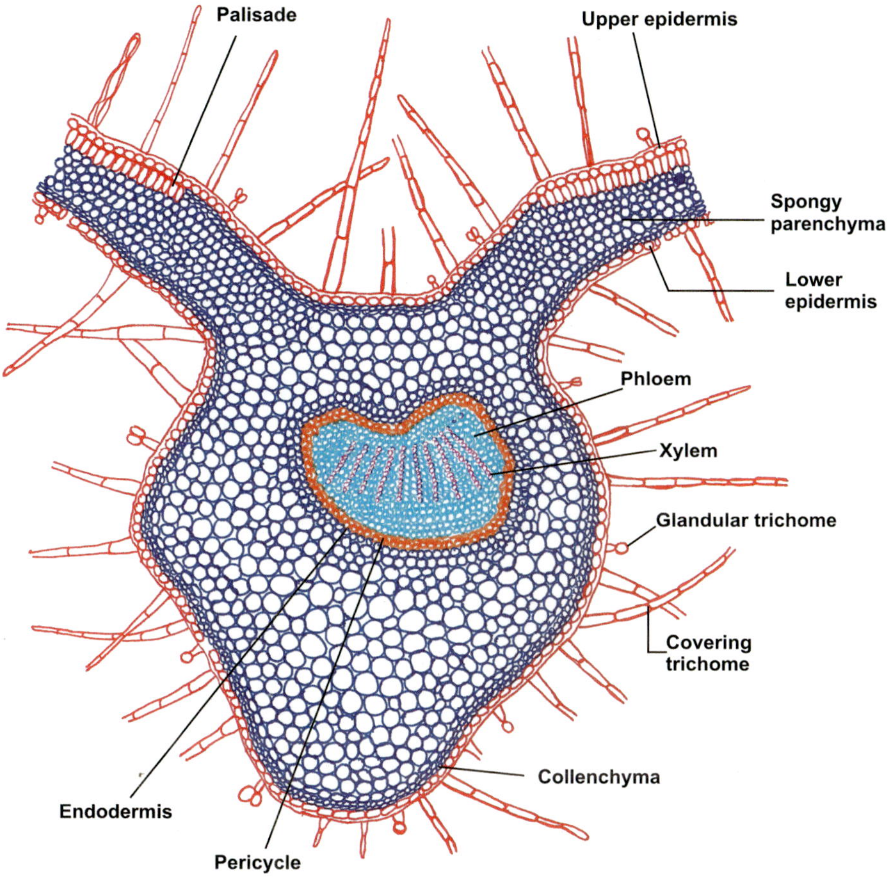

T. S. of *Digitalis purpurea* leaf

Powder Character:

The powder of digitalis leaves is dark to pale greenish in colour with a slight characteristic odour and bitter taste. Diagnostic and other characters are as follows:

a. Fragments of lamina:
   Upper epidermis is composed of irregularly shaped cells with slightly thick walls which sometimes show slight beading and pitting. Stomata are few on the upper epidermis, but numerous on the lower epidermis. Palisade cells are large and loosely packed. Lower epidermal cells are smaller with wavy walls. Cicatrices where trichomes are attached are seen on both epidermises. Powder or fragments of tissues show no calcium oxalate crystals of any type.
b. Trichomes:
   Both covering and glandular trichomes are present, seen as scattered or attached to fragments of the epidermis. Covering trichomes are numerous, long, uniseriate with three to five cells and blunt tips and finely warty. One or more of the cells may be collapsed. Glandular trichomes are few and have a single-celled stalk and bicellular (or rarely unicellular) head or a uniseriate multicellular stalk and a unicellular head.
c. Parenchyma:
   Fragments of thin-walled parenchyma are seen composed of longitudinally elongated cells.
d. Fragments of epidermis and underlying tissues are seen.
e. Pieces of veins and petioles show fibrovascular elements.
f. Vessels with reticulate, annular or spiral thickenings or with pores are observed.
g. Palisade and spongy parenchyma with chloroplasts are also seen.
h. Stone cells and calcium oxalate crystals are absent.

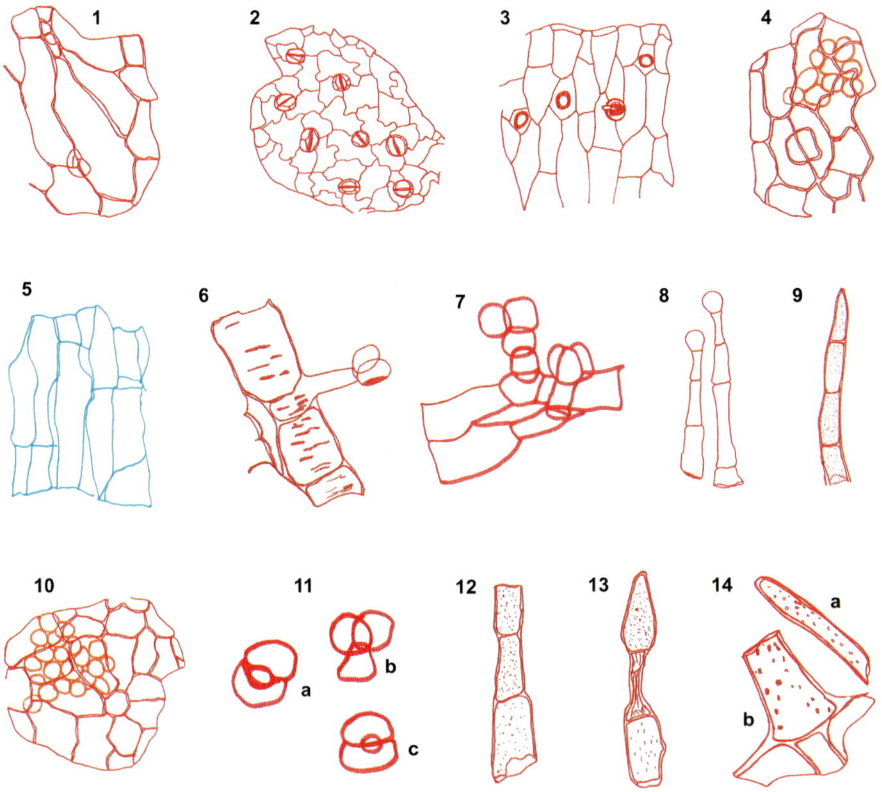

1. Epidermis in surface view showing cicatrices
2. Lower epidermis in surface view with anomocytic stomata
3. Epidermis from over a vein in surface view, showing cicatrices
4. Upper epidermis in surface view showing a cicatrix and underlying palisade cell
5. Cortical parenchyma in longitudinai view
6. Epidermis in sectional view showing pitting in the walls and a glandular trichom
7. Glandular trichomes attached to a fragment of the epidermis
8. Glandular trichomes wfth uniseriate stalks and unicellular heads
9. Part of a covering trichome
10. Upper epidermis in surface view with underlying palisade cells
11. Glandular trichomes seen a) from below b) from the side and c) from above
12. Fragment of a large covering trichome
13. Part of a covering trichome showing a collapsed cell
14. Fragments of covering trichomes : a) apical cell and
b) basal cell attached to a fragment of epidermis

16 Ephedra (*Ephedra gerardiana*)

Synonyms:

Mahuang, Som

Biological Source:

It consists of dried young stems of the plant *Ephedra gerardiana* Wall. and *Ephedra nebrodensis* (Tineo.) Stapf.

Family:

Ephedraceae

Microscopy:

The transverse section of the ephedra stem (when taken through the internode) appears more or less circular. The margin appears wavy due to ridges. The following tissues are observed from the periphery towards the centre:

A. Epidermis:
   It is composed of a single layer of thick-walled, quadrangular cells with thick and smooth cuticle. Vertical rows of sunken stomata are present between many vertical ridges of the stem. Papillae are also present in the ridges. Below the ridges, groups of non-lignified fibres are observed.
B. Cortex:
   It is composed of two to three layers of chlorenchyma (loosely arranged parenchymatous palisade cells containing chloroplasts) with outer layers of radially elongated cells and inner zone of spongy parenchyma. Cortex shows lignified as well as non-lignified fibres.

   Unlignified fibres appear like a bunch of grapes and occur below the ridges where no palisade cells are seen. Lignified fibres are found scattered, isolated or in groups of two to four. These occur in the inner layers of oval, cortical parenchyma which show chloroplasts.
C. Pericyclic fibres:
   Pericycle consists of groups of lignified fibres outside the phloem.
D. Vascular bundles:
   These are around six to ten in number radially arranged in the cortex. These are collateral, conjoint and open. Phloem is towards the outer side and appears distinctly. It contains sieve tubes and companion cells. Xylem is well developed consisting of vessels, tracheids, fibro-tracheids and parenchyma. Xylem from a mature stem shows a well-developed continuous band.
E. Pith:
   It is composed of large, thin-walled, lignified and polygonal parenchyma with intercellular spaces. Some cells contain brownish, mucilaginous masses.

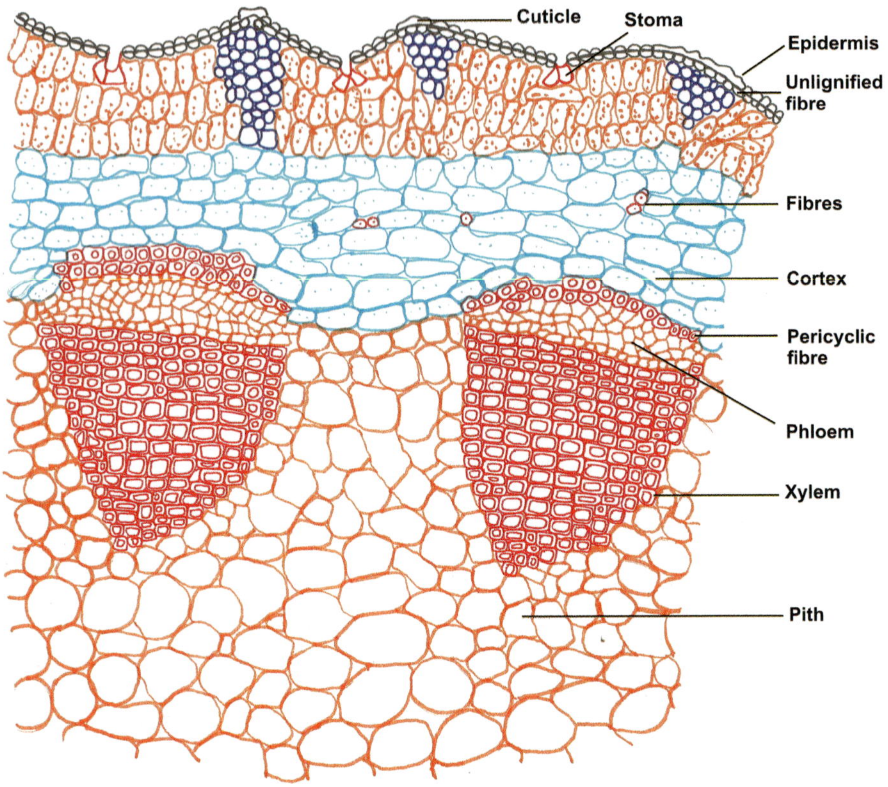

T. S. of Ephedra stem

Powder Character:

The powder of ephedra is pale yellowish brown with faint odour and slightly bitter taste. It shows the following characters microscopically:

a. Epidermal cells:
   Entire cells and fragments of cells are both observed. Cells are rounded to quadrangular with thick-ridged outer walls, sunken stomata and papillae.
b. Fibres:
   Lignified and non-lignified fibres appear, which are of uniform thickness, long, slender and cylindrical (like glass rods). Entire fibres or fragments of fibres are seen.
c. Wood elements (xylem):
   It consists of tracheids only with bordered pits.
d. Brownish matter:
   It originates from pith. It is abundant, mucilaginous and of regular shape and form.

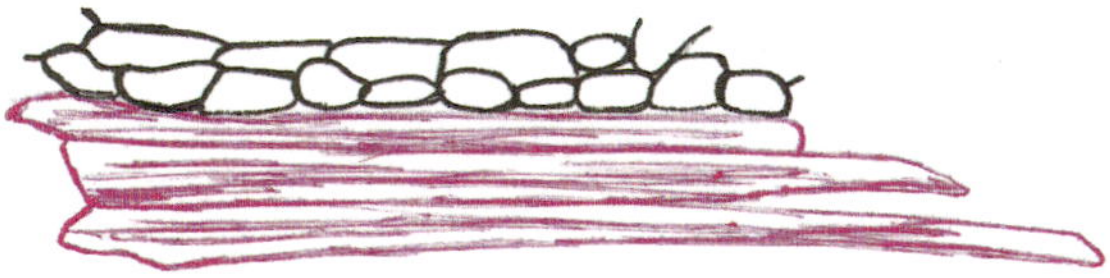

Fibres

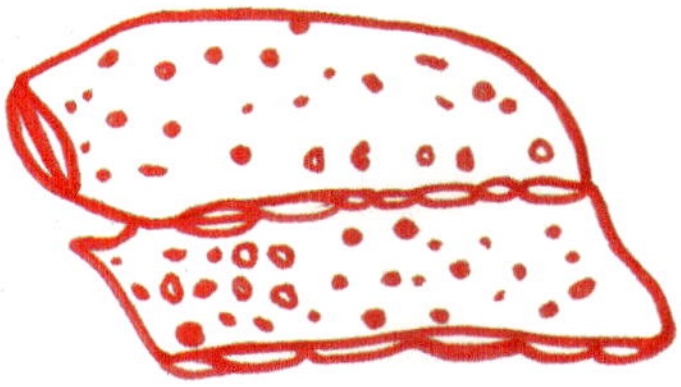

Fragment of vessels

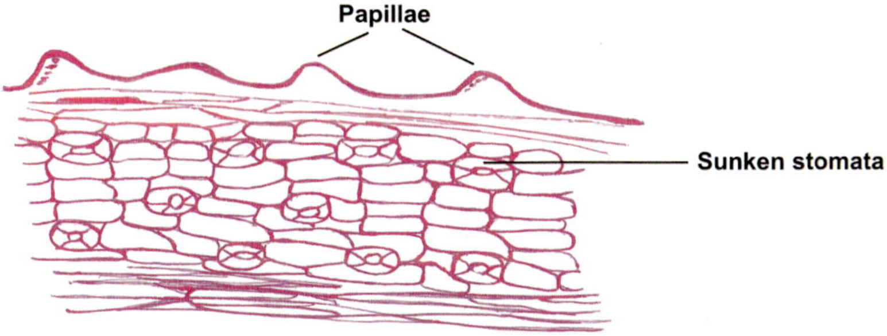

Epidermis in surface view

17 Eucalyptus (*Eucalyptus globulus*)

Synonyms:

Nilgiri, Tailpatra

Biological Source:

It consists of fresh leaves of the plant *Eucalyptus globulus* Labill.

Family:

Myrtaceae

Microscopy:

Microscopically the transverse section of eucalyptus leaf shows isobilateral structure. The following tissues are seen in the lamina and midrib region:

A. Lamina:

i. Upper epidermis:
It consists of a single layer of small rectangular parenchymatous cells with straight anticlinal walls and thick cuticle. Trichomes are not observed. Numerous sunken stomata of anomocytic type are seen.

ii. Mesophyll:
It can be differentiated as palisade and spongy parenchyma:

a. Upper palisade:
It consists of two to five layers of thin-walled, compactly packed, small palisade cells. Calcium oxalate cluster crystals are seen. This region often shows large subglobular to ovoid schizogenous oil glands opening towards the epidermis. The oil glands, those that have discharged their contents through a split between the modified cells of the epidermis, are found to be lined with a layer of cork. This area is named as ‘cork tumour’ and appears brownish in colour.

b. Spongy parenchyma:
It is composed of three to five layers of loosely arranged irregular parenchymatous cells. These cells are projected in the direction of the palisade cells. Prisms of calcium oxalate occur near the fibres of the veins. Clusters of calcium oxalate are seen in the palisade and spongy parenchyma tissues.

iii. Lower palisade:
It resembles the upper palisade.

iv. Lower epidermis:
It is identical to the upper epidermis.

B. Midrib:
The midrib shows uniform dorsal and ventral surfaces. It appears as slightly biconvex. Epidermises continue with the lamina.

The following tissues are also observed in the midrib region:

i. Collenchyma:
Two to three rows of thick-walled cellulosic cells are observed below the upper and above the lower epidermis.

ii. Vascular bundles:
It occupies the major portion of the midrib. It appears in the centre and is arc shaped. Two small vascular bundles are also seen towards the dorsal surface. All these vascular bundles are further covered partly by a sheath of patches of lignified pericyclic fibres. Prisms and sphaeraphides of calcium oxalate are seen in the midrib region.

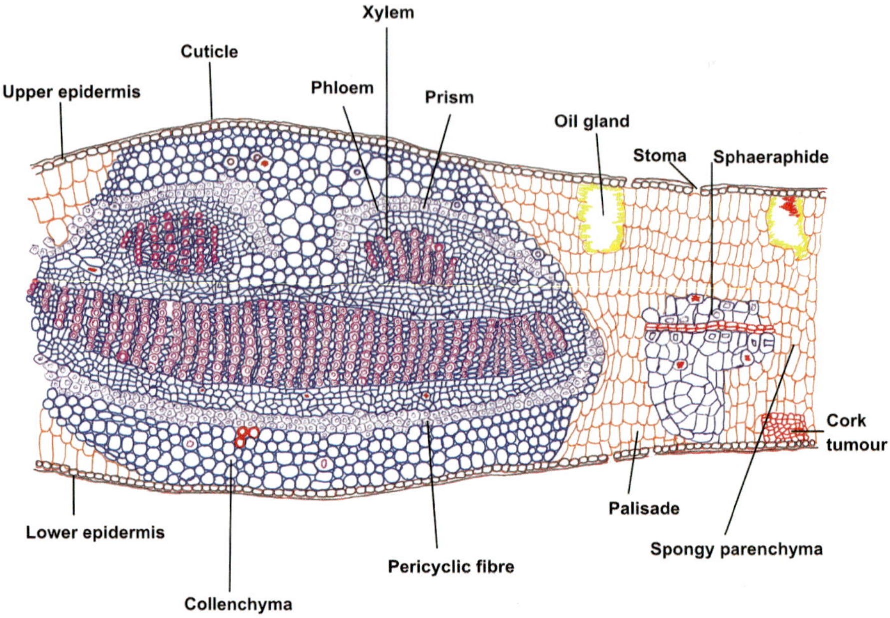

T. S. of Eucalyptus leaf

Powder Character:

The powder of eucalyptus leaves is yellowish green to light brown in colour with aromatic, characteristic odour and slightly pungent taste which gives a cooling sensation in the oral cavity. The following characters are seen microscopically:

a. Epidermis:
   The fragments of epidermis as well as entire cells can be observed. Cells appear as polygonal, parenchymatous and straight walled with thick cuticle. Stomata may appear but trichomes are absent.
b. Stomata:
   Numerous anomocytic stomata are seen on both upper and lower epidermises. These are numerous, prominent, sunken and well developed.
c. Calcium oxalate crystals:
   Rosettes and prismatic crystals of calcium oxalate are observed as scattered in the powder.
d. Pericyclic fibres:
   These are lignified, isolated or in small groups and have tapering ends.
e. Xylem vessels:
   Numerous fragments of spiral, annular or pitted vessels are seen.
f. Cell contents:
   Subglobular oil glands, oil drops and brownish cell contents can be observed.

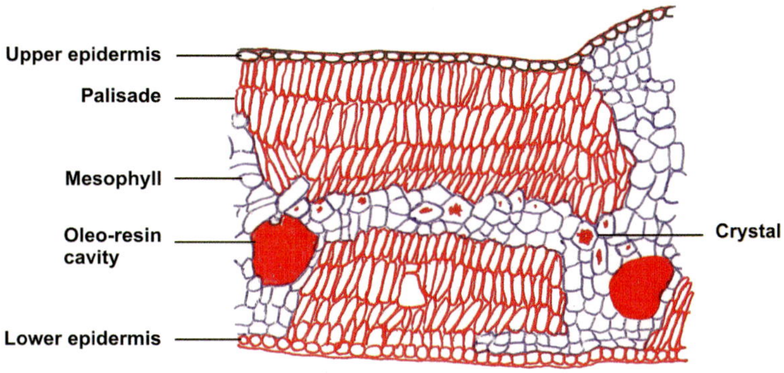

T. S. of lamina

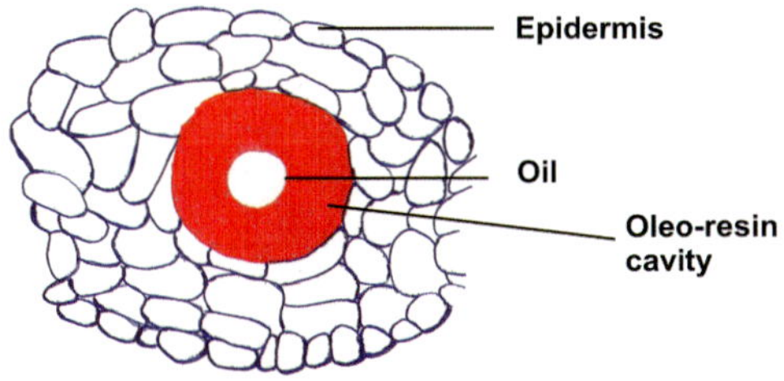

Portion of tissues from the petiole with oleo-resin cavity

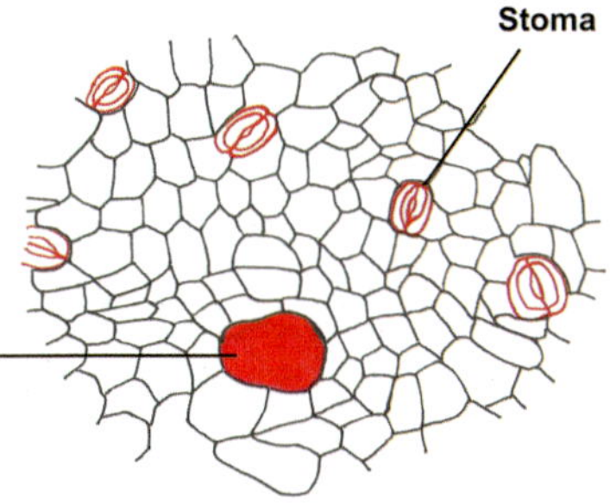

Epidermis in surface view

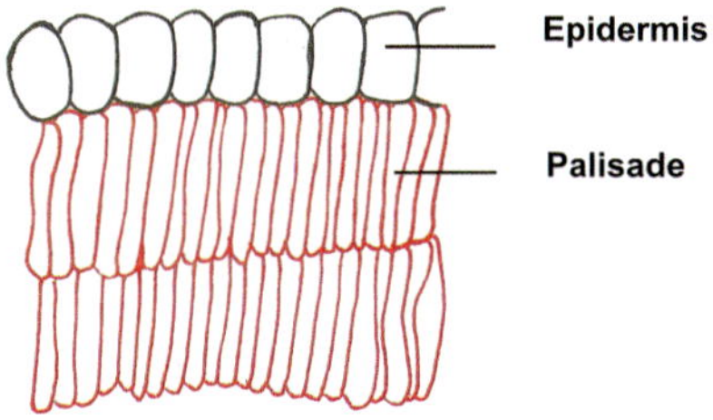

T. S. of leaf fragment

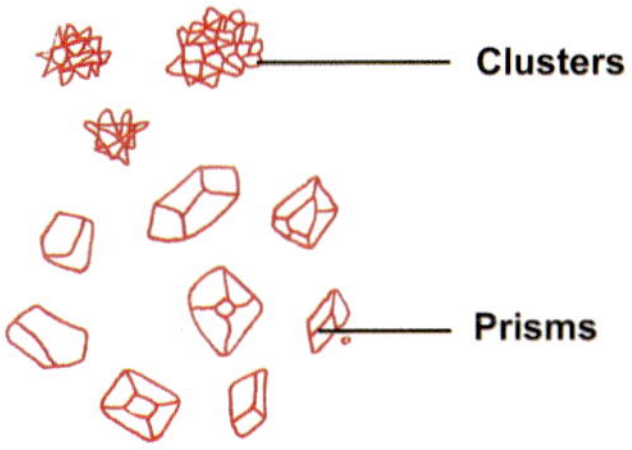

Calcium oxalate crystals

18 Fennel (*Foeniculum vulgare*)

Synonyms:

Fennel fruits, Foeniculum, Saunf

Biological Source:

It consists of dried ripe fruits of the plant *Foeniculum vulgare* Miller.

Family:

Umbellifereae

Microscopy:

Fennel shows features of a typical umbelliferous fruit.

Cremocarp:
Cremocarp is a variety of schizocarp (splitting fruit) which divides into two one-seeded portions, each corresponding to one carpel. This carpel itself does not open to liberate the seed; thus, these schizocarps are indehiscent fruits. A cremocarp consists of two parts, each of which is called a 'mericarp'. These two mericarps are connected by a thick-walled sclerenchymatous central stalk called 'carpophore'. A single seed is seen in each mericarp. Raphe is a single ridge of vascular bundle at the middle of the commissural surface. The carpophore is situated just in front of the raphe.

Transverse section of a mericarp shows two prominent surfaces: commissural and dorsal. The commissural surface is flat with two distinct ridges and carpophore in the middle. The dorsal surface shows three ridges. Therefore, each mericarp shows five primary ridges. A mericarp can be divided into pericarp and seed.

A. Pericarp:
   It is made up of epicarp, mesocarp and endocarp:

   i. Epicarp (exocarp):
      The epicarp of the pericarp is also called epidermis. It surrounds the entire fruit and consists of a layer of polygonal, tabular, tangentially elongated cells with nonstriated and smooth cuticle and shows occasional stomata.
   ii. Mesocarp:
      The bulk of the mesocarp is parenchymatous. It contains much thickened and lignified parenchyma in the region of the vascular strands of the ribs. These thickened walls have large, oval or rounded pits; the bands of thickening between them give a reticulate appearance to the walls. This is one of the characteristic features of the fennel fruit.

      Mesocarp shows five vascular bundles below each ridge (thus ridges are known as primary ridges) and six vittae. Reticulate lignified parenchyma is seen around the vascular bundles.

      Vascular bundles are five in number, bicollateral and present below each ridge.

      Vittae are schizogenous oil cells, four vittae present on the dorsal side and two on the commissural surface. Vittae are about 250 μ maximum wide and taper towards the base and up to the apex of the fruit. Walls appear brown and each duct is divided into chambers by transverse partitions. Vitta is lined by an epithelium of small polygonal tabular cells. The number and position of vittae are many times characteristic of individual umbelliferous fruits. Secondary ridges occurred between the primary ridges. The number, distribution and arrangement of ridges and vittae give valuable information for identification of fruits.
   iii. Endocarp:
      It consists of narrow elongated cells having a parquetry arrangement. In the transverse section, these cells appear as long narrow rectangular cells with scattered groups of very short cells.

B. Seed:
   It consists of testa, endosperm and embryo.

   i. Testa:
      It is single layered and yellowish brown in colour.
   ii. Endosperm:
      It is thick walled, polygonal, colourless cellulosic parenchyma containing fixed oil globules, aleurone grains and rosette crystals of calcium oxalate.
   iii. Embryo:
      A crescent-shaped embryo is observed in the sections passing through the apical region of mesocarp. It is seen at the centre of the endosperm. Cells are polygonal, thick walled and parenchymatous.

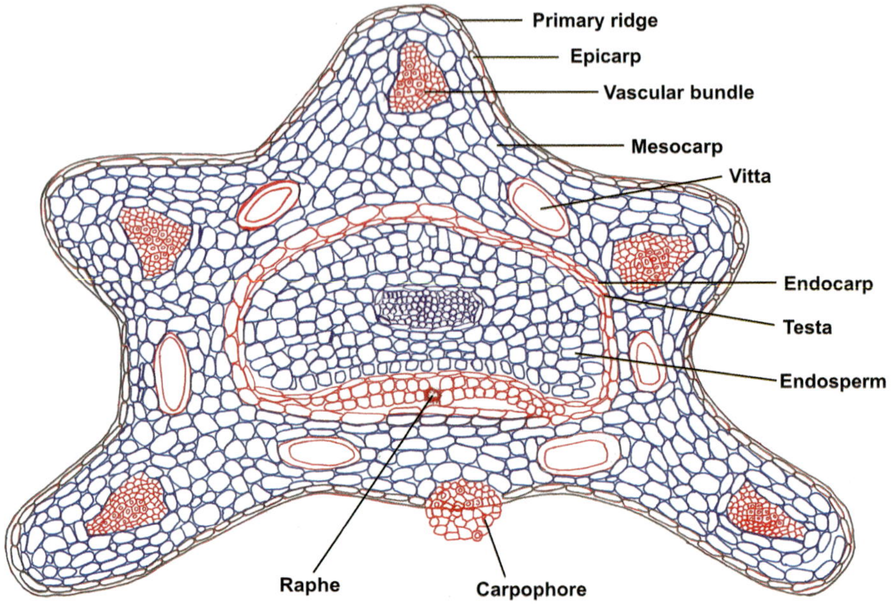

T. S. of Fennel fruit

Powder Character:

The powder of fennel is yellowish brown to greenish brown with a pleasant, aromatic odour and somewhat sweetish taste. It shows the following characters microscopically:

a. Epicarp:
   It is composed of a layer of colourless, thin-walled cells, polygonal in surface view and with a smooth cuticle. It may show a very few stomata. Stomata may be surrounded by two to four radiating cells and are found on some fragments of epicarp. Fragments of epicarp may show slight thickening and beading of the anticlinal walls.
b. Mesocarp:
   The reticulate parenchyma of the mesocarp is composed of ovoid, elongated sub-rectangular cells with thick, lignified walls with conspicuous oval or rounded pits. These parenchymatous cells of the mesocarp occur in groups and are frequently found associated with fibrovascular tissue or with fragments of the endocarp.
c. Endocarp:
   It consists of a layer of thin-walled, lignified cells, elongated in surface view and arranged in groups of about six or more cells with their long axes parallel to each other, i.e. parquetry arrangement of cells.

d. Vittae:
   Numerous yellowish-brown fragments of vittae are observed which are composed of thin-walled cells. These appear irregular in shape and are found scattered.
e. Endosperm:
   Fragments of endosperm are abundant and are composed of polygonal thick-walled cells with fixed oil globules and aleurone grains. These cells also show presence of micro-rosette crystals of calcium oxalate.
f. Innermost layer of mesocarp:
   It consists of slightly thick-walled cells, rounded to rectangular in surface view. This layer is found associated with the endocarp.
g. Fibrovascular tissue:
   Fragments of fibrovascular tissue are found with lignified small fibres, vessels, tracheids and a few large vessels with reticulate thickening. Some of the fragments are associated with reticulate parenchyma of the mesocarp.

Thicker-walled cells
of the endosperm

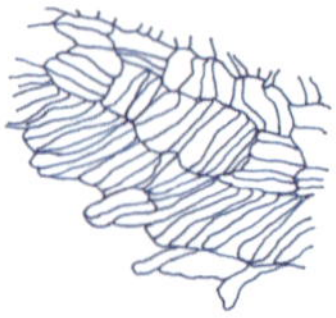

Endocarp in surface view
(Parquetry layer)

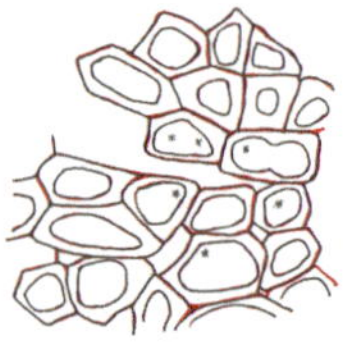

Endosperm containing
microrosette crystals of
calcium oxalate

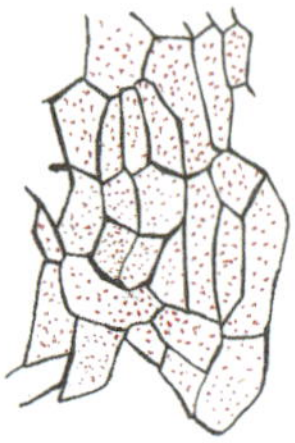

Fragment of a vitta

Part of the pericarp and seed
in sectional view showing
the reticulate parenchyma, endocarp,
testa and endosperm

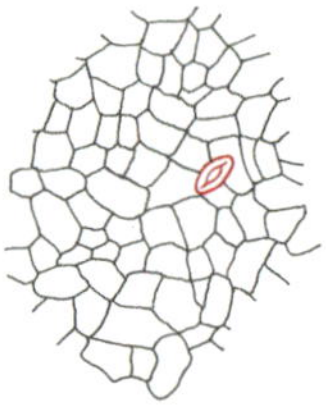

Epicarp in surface view
showing a stoma

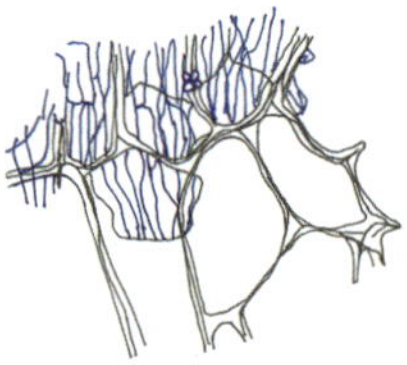

Endocarp with overlying cells of
the inner most layer of the mesocarp
in surface view

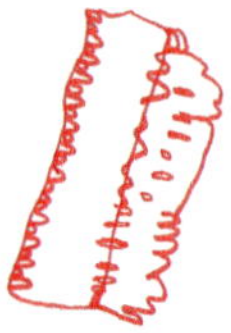

Elements from the
fibro-vascular tissue

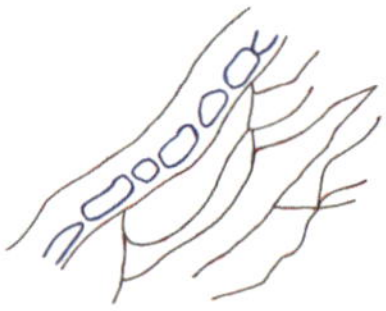

Epicarp and parenchyma
of the mesocarp in sectional view

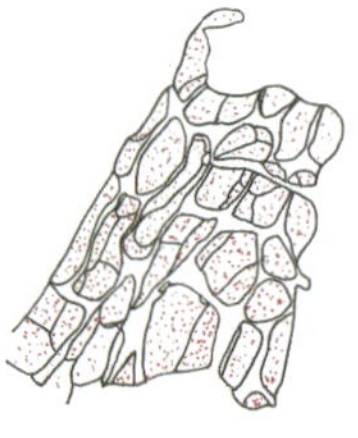

Fragment of a vitta with overlying
thick-walled cells of the innermost
layer of the mesocarp in surface view

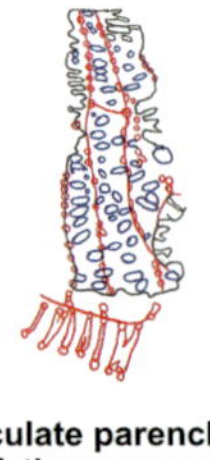

Reticulate parenchyma
of the mesocarp

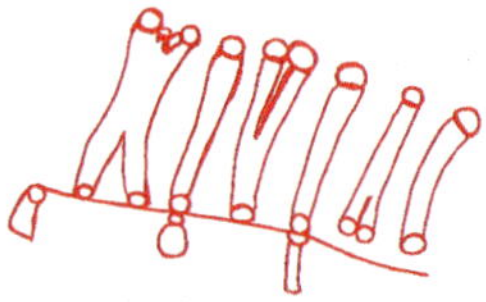

Fragment of a reticulately
thickened vessel

19 Ginger (*Zingiber officinale*)

Synonyms:

Zingiber, *Amomum zingiber* L.

Biological Source:

It consists of peeled or unpeeled rhizomes of the plant *Zingiber officinale* Roscoe.

Family:

Zingiberaceae

Microscopy:

A cross section of an unpeeled rhizome of ginger shows the following tissues from the periphery to the centre:

A. Cork:
   It can be differentiated into two zones as outer zone and inner zone:

   i. Outer zone:
      It is dark brown in colour and made up of few layers of irregularly arranged parenchymatous cells formed by suberisation of cortical cells without division.
   ii. Inner zone:
      It is few layered and made up of colourless parenchymatous cells radially arranged in regular rows and produced by the tangential division of the cortical cells.

The cork cambium is not differentiated separately.

B. Cortex:
It is broad and composed of an outer zone of flattened parenchyma and an inner zone of regular parenchyma. Cortical cells are isodiametric, thin walled with intercellular spaces. Cortex also shows scattered vascular strands and idioblasts. Some of these cells contain starch grains which are flattened and oval-oblong. Starch grains are characteristic, sac-shaped, simple, rarely compound and have terminal beaklike projection (protuberance) in which eccentric hilum is situated. Striations are distinct and prominent and run across the grains at right angles to their long axes. Amongst the starch-containing cells, there are cells which contain yellow to reddish-brown masses of oleoresin and having cuticularised walls (oleoresin cells).

Cortex contains about three rings of collateral, conjoint and closed vascular bundles in the inner zone. Vascular bundles which are at the periphery of the ground tissue are not fibrovascular bundles. A group of sclerenchymatous fibres partially covers the vascular bundles of the cortex usually in arc shape and sometimes completely surrounds the bundle. The fibres are lignified, thin walled, sometimes non-lignified in small portions and pitted and have delicate transverse septa. Xylem vessels are annular and spiral or reticulate and have unlignified thickenings. Many of the vessels are separately associated with a slender and elongated cell containing brown pigment. Phloem shows well-developed sieve tubes.

C. Endodermis:
It is single layered and made up of thick, radial-walled cells, without any starch.

D. Ground tissue:
It is made up of large, parenchymatous cells containing abundant starch, oleoresin cells and vascular bundles. The outermost single layer of the ground tissue is pericycle, and a ring of small vascular bundles is arranged in the parenchyma just below the pericycle. These vascular bundles are not covered with fibres. The remaining part of the ground tissue contains fibrovascular bundles, starch grains and numerous oleoresin cells. Starch grains are flattened and rectangular to ovate with hilum situated in a terminal projection and also show prominent, transverse striations. Vascular bundles are similar to the fibrovascular bundles of the cortex.

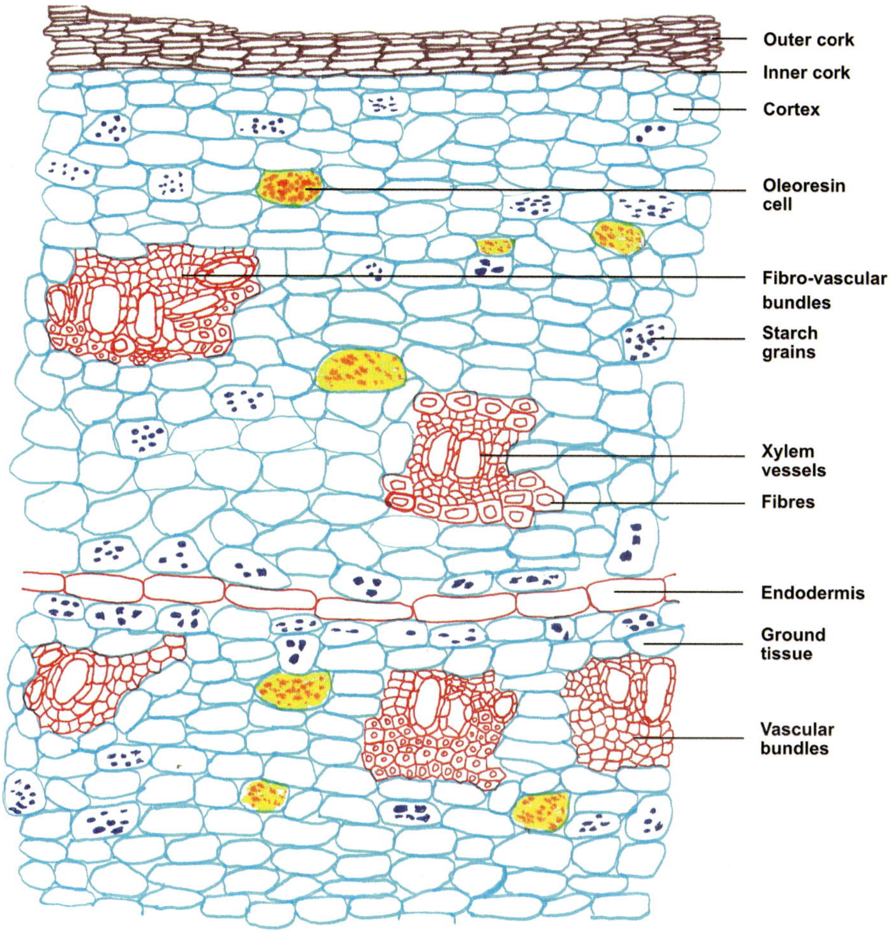

T. S. of Ginger rhizome

Powder Character:

The powder of ginger is pale yellowish cream coloured with aromatic, characteristic, pleasant odour and characteristic, pungent taste.

It shows the following characters microscopically:

a. Starch grains:
These are characteristic and abundantly present. These are mostly simple, large, flattened and oblong to sub-rectangular or sac shaped with a distinct eccentric hilum. Faint transverse striations are observed. Compound granules with two components also occur rarely.
b. Fibres:
Fibres usually occur in groups and may be also found associated with vessels. These are large; walls are thin, pitted and slightly lignified. Very thin transverse septa occur at intervals. Fibres may give a faint reaction for lignin.
c. Vessels:
These are large and occur in small groups associated with the fibres. These are reticulately thickened and frequently show distinct, regularly arranged, rectangular pits. These vessels show narrow, thin-walled cells with dark brown pigment. A few small, spirally or annually thickened vessels also occur. Vessels give faint reaction for lignin.
d. Oleoresin cells:
These cells are thick walled and occur as yellowish and ovoid to spherical and either singly or in small groups in the parenchyma.
e. Parenchyma:
These cells are numerous, thin walled and round to oval in outline with small intercellular spaces. Idioblasts contain yellow–brown oleoresin or starch granules. Walls of the many parenchymatous cells are characteristically wrinkled. Groups of parenchyma are also found associated with thin-walled tissue composed of many rows of collapsed cells.

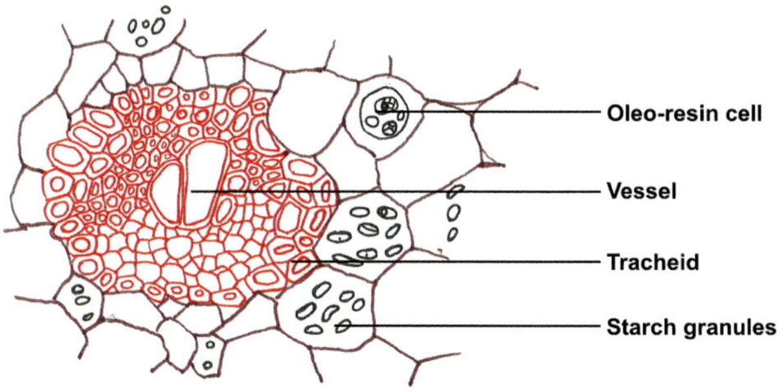

T. S. through xylem portion

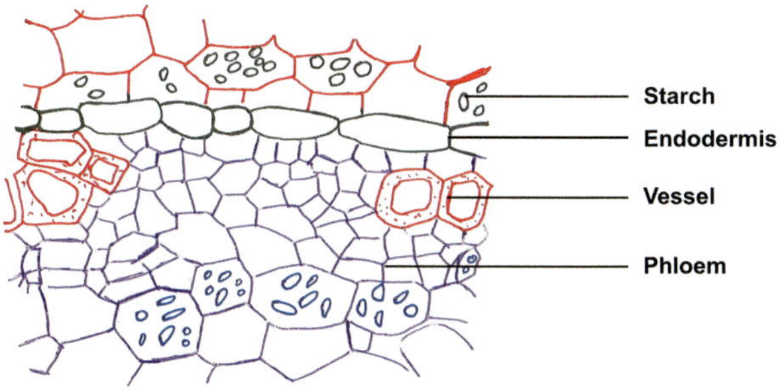

T. S. of endodermis and xylem portion

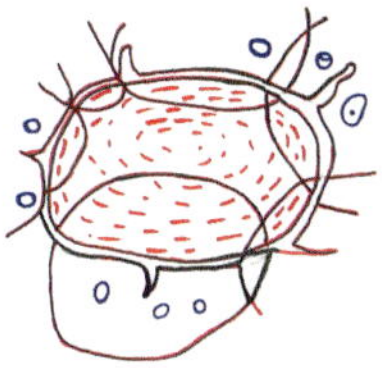

Oleo-resin cell

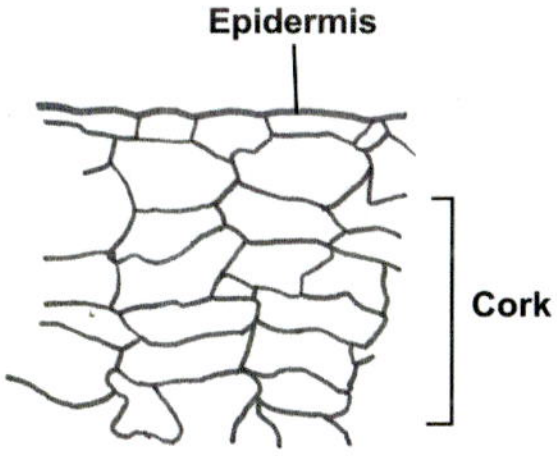

T. S. of outer rhizome

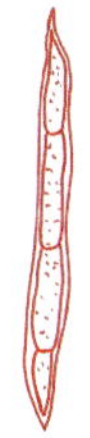

Fibre

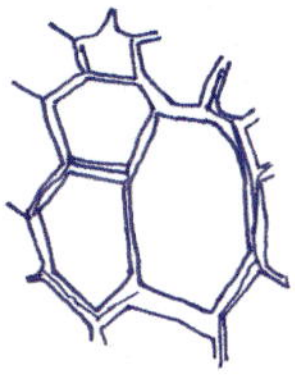

Cork cells

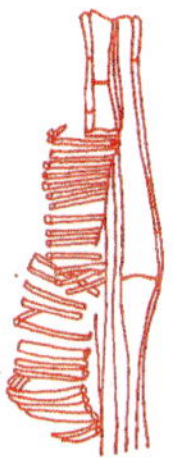

Septate fibres with attached vessels

20 Ipecacuanha (*Cephaelis ipecacuanha*)

Synonyms:

Ipecac, Cartagena, Nicaragua, *Panama ipecacuanha*

Biological Source:

It consists of dried rhizomes and roots of the plant *Cephaelis ipecacuanha* (Brot.) A. Rich. or *Cephaelis acuminate* Karsten.

Family:

Rubiaceae

Microscopy:

The transverse section of the ipecac root shows a circular outline.

The following tissues can be observed in the transverse section microscopically:

A. Periderm:
   Cork and phellogen collectively form the periderm:

   i. Cork:
      It is made up of three to five layers of thin, narrow, brown and tangentially elongated to rectangular isodiametric cells, filled with dark brown granular material.
   ii. Phellogen:
      It is seen immediately below the cork. It is composed of two to three rows of tangentially elongated thin-walled cells. A few of these cells show presence of starch granules.

B. Cortex:
The secondary cortex (phelloderm) may be observed above the cortex. The cortex is composed of many layers of thin-walled cellulosic parenchyma with small intercellular spaces. The cortical parenchyma contains acicular raphides of calcium oxalate either in bundles (as idioblasts) or scattered all over. Starch grains are abundantly present. These are simple, single or compound mostly with two to four grains or may grow up to eight grains. The simple grains are less in number. The individual starch grains appear muller shaped.

C. Phloem:
It is totally parenchymatous and seen as many patches of small groups of perforated sieve tubes around well-developed xylem. Sclerenchymatous fibres or cells are not observed.

D. Cambium:
It is seen sometimes as two to three layers of flattened cells.

E. Xylem:
Xylem is entirely lignified and occupies the major (about one third) portion of the transverse section. It is transversed regularly by lignified medullary rays. Each cell of medullary ray is radially elongated and contains starch. The compact, dense, central mass of xylem (secondary xylem) consists of tracheids, tracheidal vessels, xylem parenchyma, xylem fibres and substitute fibres. Tracheids have pitted walls. Vessels are not easily distinguished from tracheids. Xylem parenchyma cells are packed with starch grains which are also in xylem fibres.

Transverse section of an ipecac rhizome shows all these tissues along with a ring of xylem and large pith. The pericycle contains characteristic sclerenchymatous cells. Protoxylem shows spiral vessels. Pith is composed of pitted and somewhat lignified parenchyma.

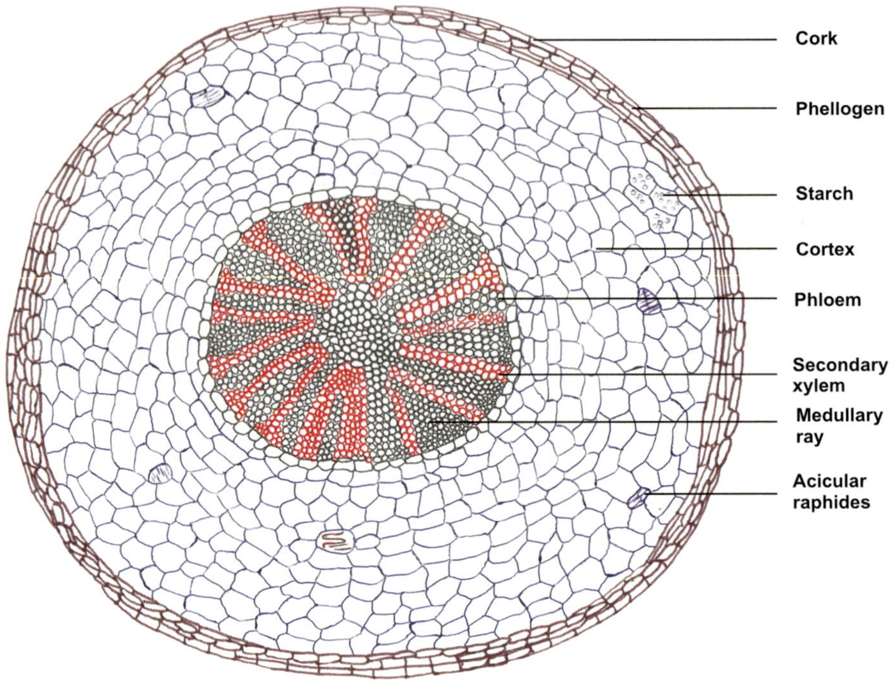

T. S. of Ipecac root

Powder Character:

Powdered ipecac is light greyish-fawn powder with a slight odour and bitter taste. It is stimulatory and irritant to mucous membrane.

It shows the following characters microscopically:

a. Fragments of cork:
   These are seen abundantly. Cork is reddish brown, composed of many layers of thin-walled polygonal isodiametric cells with slightly lignified walls.
b. Starch granules:
   These are single and compound, mostly with two to eight components. Individual granules are small and spherical to ovoid in shape. Few grains show prominent round, pointed or cleft-shaped hilum.
c. Calcium oxalate crystals:
   Idioblasts with acicular raphides of calcium oxalate are observed. These crystals are either in bundles or found scattered singly in the powder.
d. Parenchyma of the xylem:
   It is found abundant in number. Cells are rectangular, thick walled and longitudinally elongated with scattered border or simple pits.

e. Tracheids:
   Tracheids and tracheidal vessels are small, lignified and thick walled with small bordered pits and found in groups.
f. Fibrous cells of xylem:
   These occur as single or associated with other xylem elements. These are elongated with tapering ends. The lumen is divided by thin transverse septa. Walls are thick and lignified and show small pits.
g. Parenchyma of the phelloderm:
   It is abundant in number and filled with starch granules (idioblasts). The cells are thin walled and round or oval with small intercellular spaces. Few fragments of parenchyma of pith of the rhizome are large and slightly thick. Cells are lignified and have many simple pits.
h. Sclereids:
   These are from rhizomes, found singly or in small groups. These are large and rectangular, with moderately and unevenly thickened walls, and have numerous large, conspicuous pits.

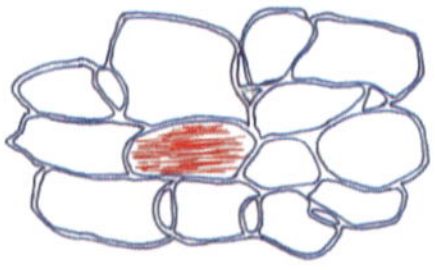

Parenchymatous cells of the phelloderm showing a bundle of acicular crystals of calcium oxalate in one of the cell

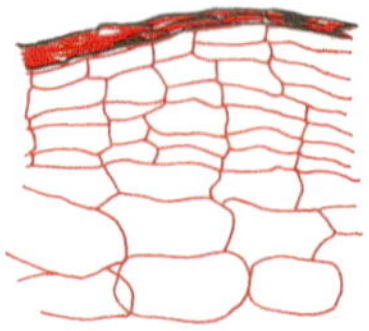

Part of the cork and phelloderm in sectional view

Fragment of xylem elements

Part of a fibrous cell showing septa

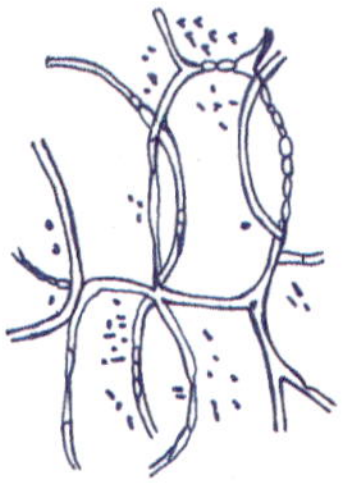

Parenchyma of the pith of the rhizome

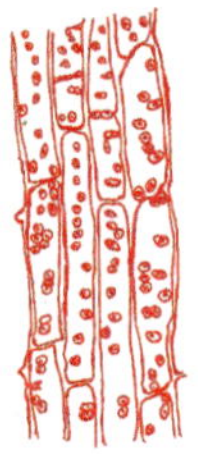

Lignified xylem parenchyma in longitudinul section

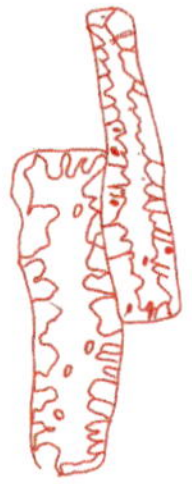

Sclereids from the rhizome

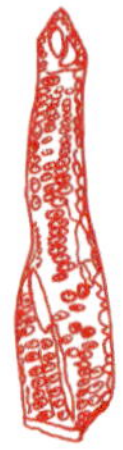

Tracheidal vessels

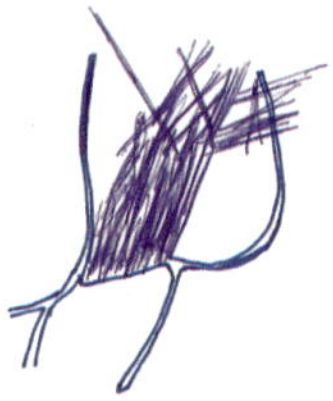

Parenchyma with acicular crystals of calcium oxalate

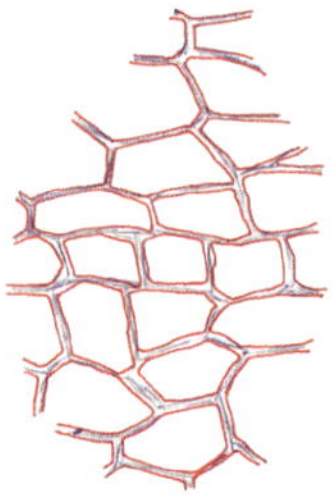

Cork in surface view

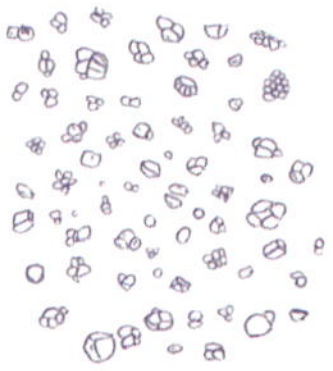

Starch granules

Part of a group of tracheids

21 Ispaghula (*Plantago ovata*)

Synonyms:

Ispaghula, Isabgul, Indian psyllium, Isabgol, Flea seed

Biological Source:

It consists of dried ripe seeds of the plant *Plantago ovata* Forsk.

Family:

Plantaginaceae

Microscopy:

The transverse section of ispaghula seed shows a crablike appearance and possesses two surfaces: a convex and a concave.

The following tissues are seen in the transverse section microscopically:

A. Testa:
   Epidermis and pigment layer form the testa of the seed:

   i. Epidermis:
      It is single layered and composed of polygonal, tabular, thin-walled, colourless, shining cells containing mucilage abundantly. Epidermis and sometimes mucilage show presence of starch grains, simple or compound. Epidermis is not present on the ventral surface of the seed.
   ii. Pigment layer:
      It is a single layer and yellowish brown in colour.

B. Endosperm:
   It occupies the major portion of the seed. Cells are thick walled, polygonal and pitted containing aleurone grains and fixed oil. The cells of the outermost layer of the endosperm appear palisade like in shape.
C. Embryo:
   It contains two polyhedral cotyledons in the centre of the endosperm but slightly towards the convex surface. Cells contain aleurone grains and oil globules. The cells of the cotyledons are polyhedral and thin walled in surface view. Three to five vascular bundles are observed in each cotyledon. Cells of radicle are thin walled and uniformly arranged in layers.

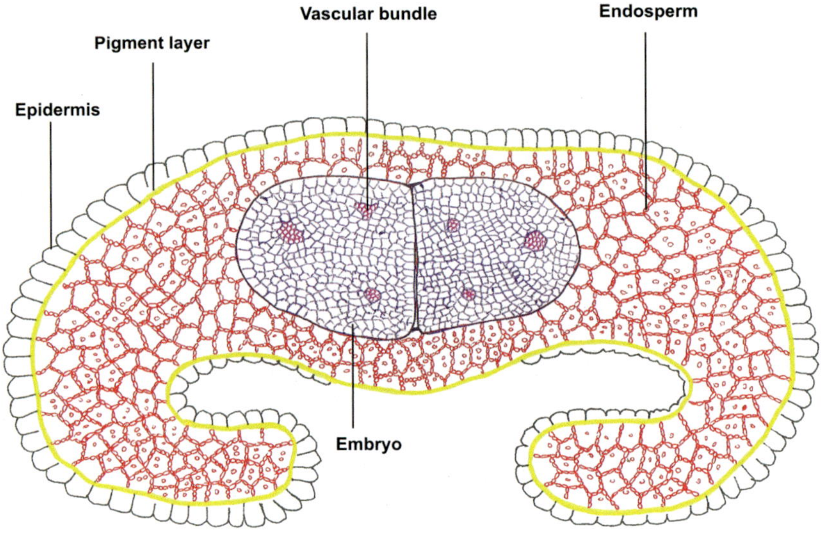

T. S. of Ispaghula seed

Powder Character:

The powder of ispaghula seeds is pale pinkish fawn in colour with a faint odour and very mucilaginous taste.

It shows the following diagnostic characters:

a. Epidermis of the testa:
   Entire cells or fragments of cells are observed. It is composed of large cells with transparent thin walls containing mucilage abundantly. These cells swell rapidly in aqueous mount and observed as slightly rounded to polygonal in surface view. When seen from above, cells appear elongated to rectangular. Swelling of cells with mucilage mainly takes place in a radial direction. Mucilage is stained with solution of ruthenium red. Occasionally small, simple or compound starch granules are seen in some of the epidermal cells and may be found as embedded in the mucilage.

b. Fragments of the endosperm:
   The endospermic cells are thick walled with many large, very conspicuous pits, usually found attached to the pigment layer. The pigment layer is the inner layer of the testa composed of indistinct, thin-walled cells with yellowish-brown pigment. Endosperm contains oil globules which show red colour with Sudan red III. Aleurone grains give yellow colour with alcoholic picric acid solution.
c. Fragments of embryo:
   These are composed of small, thin-walled cells. Cells of cotyledons are polygonal to slightly round. Fragments of radicle show thin-walled, regularly arranged, uniform cells.
d. Starch granules:
   These are seen occasionally in epidermal cells and some as embedded in the mucilage of the cells. These granules are small and simple or compound with four or more components.

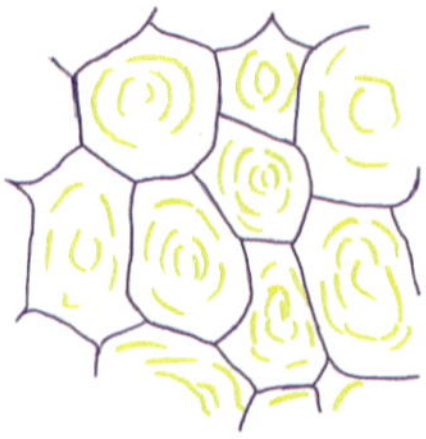

Epidermal cells of the testa in surface view, from above, showing mucilage

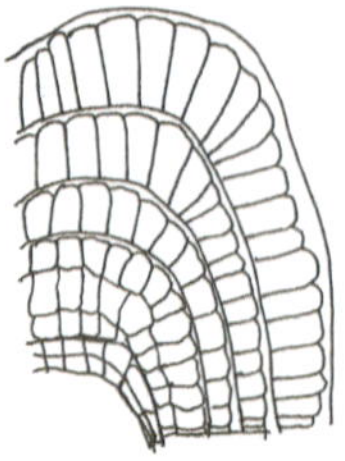

Part of the radicle of the embryo

Starch granules

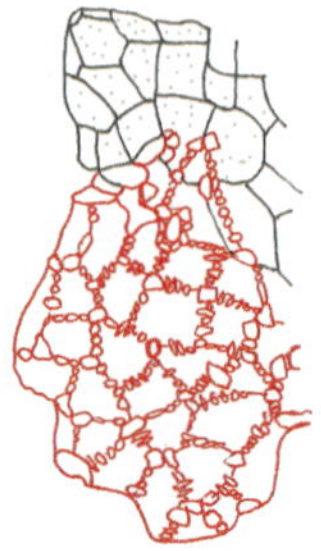

Part of the endosperm in surface view with associated layer of the testa containing pigment

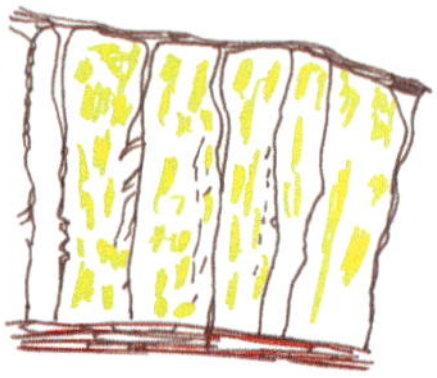

Epidermis of the testa in side view showing the swollen cells containing mucilage

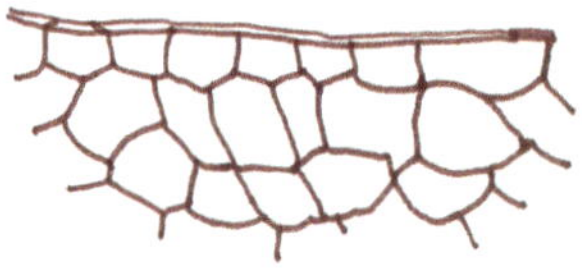

Part of a cotyledon in sectional view

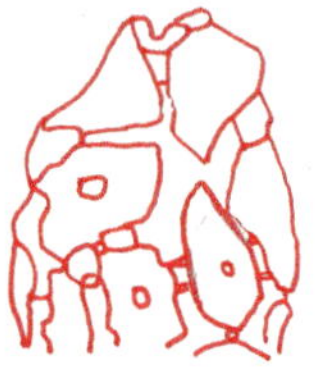

Part of the endosperm in surface view

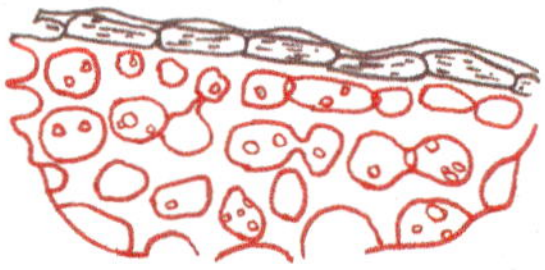

Inner layer of the testa containing pigment and part of the endosperm in sectional view

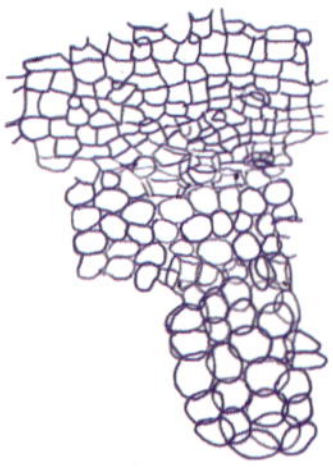

Layers of the cotyledons in surface view

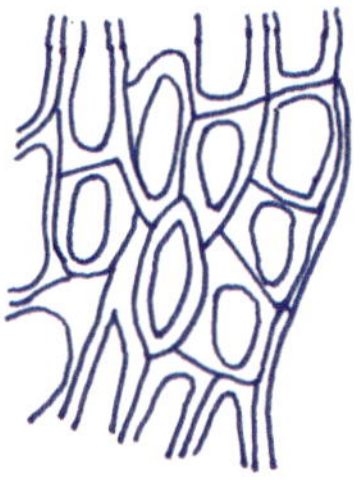

Epidermis of the testa in surface view, from below, in an aqueous mount

22 Kurchi (*Holarrhena antidysenterica*)

Synonyms:

Kurchi, Holarrhena bark

Biological Source:

It consists of dried stem bark of the plant *Holarrhena antidysenterica* Wall.

Family:

Apocynaceae

Microscopy:

Transverse section of the kurchi bark shows the following tissues microscopically:

A. Periderm
   It can be differentiated into cork, phellogen and phelloderm:
   i. Cork:
      It is composed of several (5–12) layers of thin-walled, rectangular, uniformly arranged, tangentially elongated cells. Some of these cells contain yellowish-brown matter.
   ii. Phellogen:
      It is made up of two layers of colourless, thin-walled, tangentially elongated rectangular cells.
   iii. Phelloderm:
      It is composed of a few (up to 10) layers of thin-walled, irregular polygonal to rectangular parenchymatous cells arranged in radial rows. These cells show prismatic calcium oxalate crystals and a few simple starch grains within them.

B. Cortex:
   It is a broad zone of parenchymatous, polygonal, thin-walled irregular cells interspersed with groups of lignified, pitted stone cells. These stone cells (sclereids) have a large lumen and are of various shapes (rectangular, oval or elongated) and sizes. Individual sclereid cell is more or less rounded to oval and thick walled with numerous pits. The groups of stone cells sometimes form a continuous band. The cortical parenchyma surrounding the stone cells and stone cells containing rhomboidal crystals are characteristic features of kurchi bark. Starch grains also occur in the cortical parenchyma. A few non-lignified pericyclic fibres are also seen sometimes. A zone of sclereids containing calcium oxalate crystals alternating with parenchymatous zone indicates the limit of the cortex.
C. Secondary phloem:
   Below the cortex, secondary phloem region starts which mainly comprises of phloem parenchyma, sieve tubes and companion cells, medullary rays and stone cells. The phloem parenchyma is similar to the cortex, transversed longitudinally by medullary rays at regular intervals. Medullary rays are seen as if not arranged uniformly and appear to run in different directions. These rays are one to three seriate, almost straight, narrow and wide towards the outside and consist of thin-walled, radially elongated parenchymatous cells. Phloem parenchyma and medullary ray cells contain starch grains. Phloem fibres are absent. Stone cells are arranged in tangential rows and are separated by medullary rays. The stone cells in the secondary phloem region also are encircled by a sheath of parenchyma containing rhomboidal crystals of calcium oxalate.

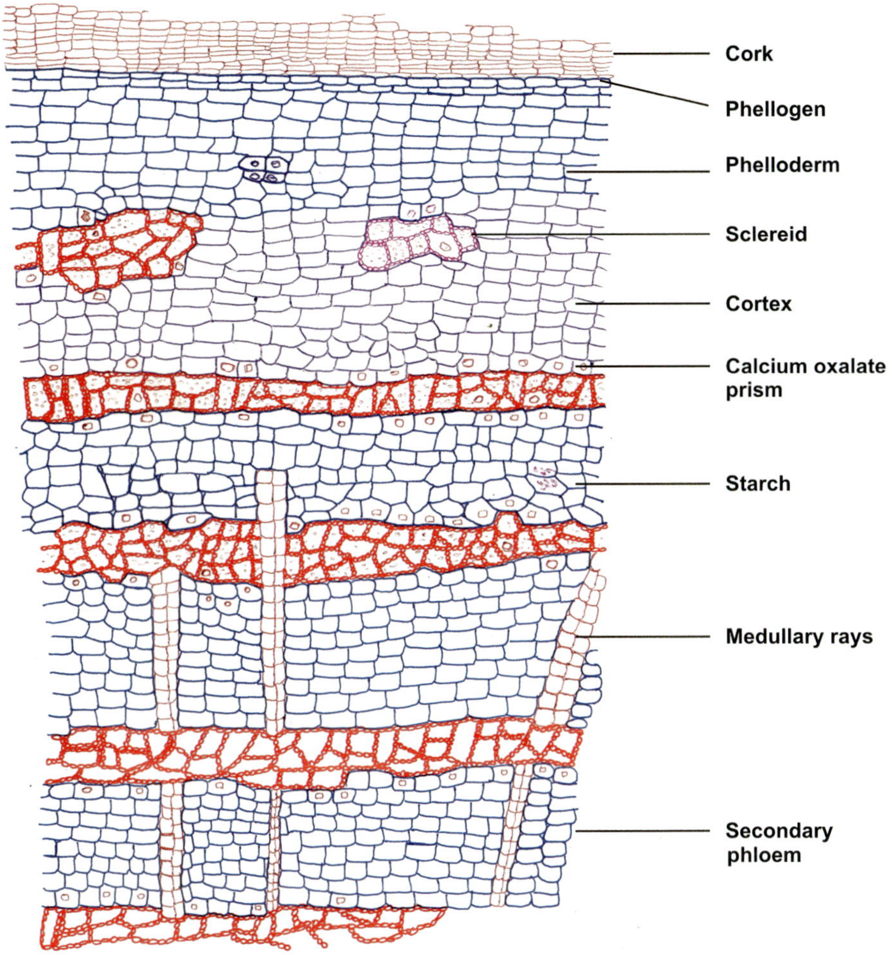

T. S. of Kurchi bark

Powder Character:

The powder of kurchi is light brown in colour with slight odour and bitter taste. It shows the following characters microscopically:

a. Cork cells:
   These cells are thin walled, some colourless and others with yellowish-brown matter.
b. Stone cells:
   These cells are observed as intact and as fragments. These are lignified and observed in groups. Individual cell appears rectangular to elongated, with straight walls and pitted thickening. A few cells contain calcium oxalate prisms. Groups of stone cells are surrounded by sheath of parenchymatous cells containing prisms of calcium oxalate.
c. Calcium oxalate crystals:
   Prismatic calcium oxalate crystals are observed as scattered in the powder. These are small and uniform.
d. Starch grains:
   These are few, spherical to ovoid, simple or compound with two to four components and found scattered.
e. Medullary rays:
   Phloem parenchyma divides the medullary rays transversely at right angles, and thus these rays are seen very rarely in the powder.
f. Phloem fibres are absent in the powder.

23 Linseed (*Linum usitatissimum*)

Synonyms:

Flaxseed, Linum, Alsi

Biological Source:

It consists of dried ripe seeds of the plant *Linum usitatissimum* Linn.

Family:

Linaceae

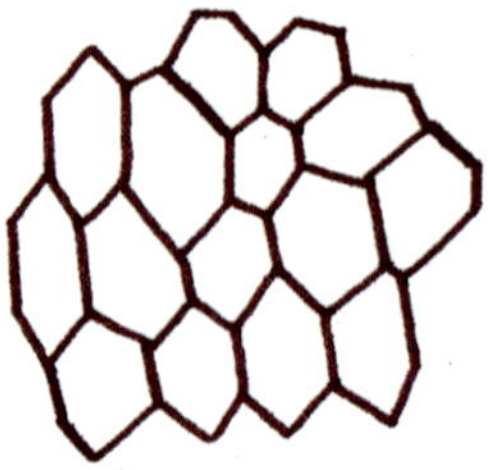

Cork in surface view

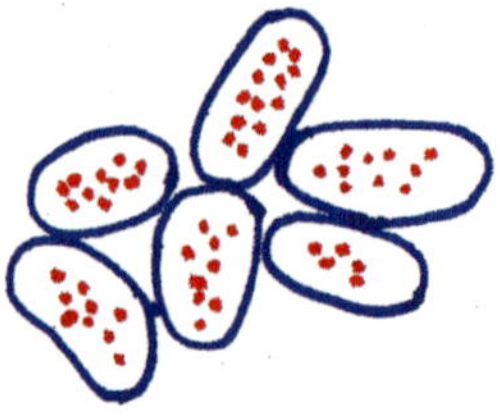

Stone cells

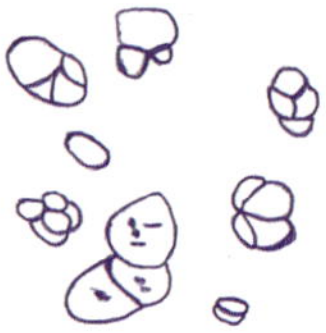

Starch granules

Fibres

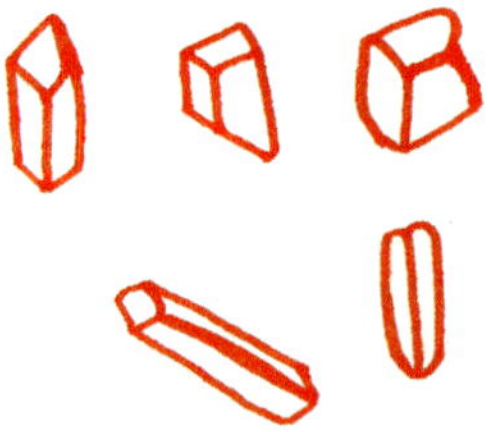

Calcium oxalate Crystals

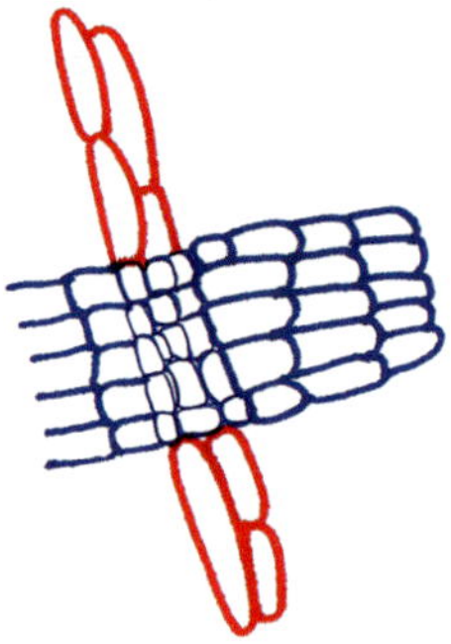

Parenchyma

Microscopy:

The transverse section of the linseed seed shows a distinct testa, a narrow endosperm and a pair of large, plano-convex cotyledons.

It shows the following tissues microscopically:

A. Testa:
   It is differentiated into outer and inner integuments:

   i. Outer integument (outer coat):
      It consists of epidermis and subepidermis:

a. Epidermis:
   It is a single layer of polygonal tabular cells with thin, shining anticlinal walls. The lumen is filled with stratified mucilage with a few starch granules embedded within. The outer cell walls become stratified when swollen in water. The inner tangential walls are suberised and the outer walls show thin cuticle. Mucilage is stained blue with iodine and red with ruthenium red.
b. Subepidermis:
   It is sometimes termed as 'round-celled layer'. It consists of one or two layers of yellowish-brown cylindrical collenchyma (round cells or radial layers) with distinct triangular intercellular air spaces. Epidermis and subepidermis together form the outer seed coat.
c. Inner integument (inner coat):
   It is made up of the following different layers:

   i. Sclerenchymatous layer:
      It is a single layer of small, longitudinally elongated lignified sclereids (stone cells) and appears as a reddish-brown layer. Cells are compactly arranged and have thick, pitted lignified walls with a very small lumen.

   ii. Parenchymatous layer (hyaline layer):
      It is made up of one or two layers of thin, tangentially elongated, collapsed parenchymatous cells. The outermost layer of hyaline is composed of narrow, elongated cells with their long axes at right angles to those of the sclereids.
   iii. Pigment layer (inner epidermis):
      It is composed of a single layer of flat, sub-rectangular to polygonal tabular cells with thickened pitted walls and containing amorphous reddish-brown contents.

B. Endosperm:
   It is narrow and thin and encircles the cotyledons. Cells of endosperm and cotyledons are parenchymatous, polygonal with thickened walls, cellulosic and colourless and contain aleurone grains and abundant oil globules. Aleurone grains have a well-developed crystalloid and globoid. A few starch grains occur in unripe seeds only.
C. Cotyledons:
   Cells and cell contents of cotyledons are similar to that of endosperm.

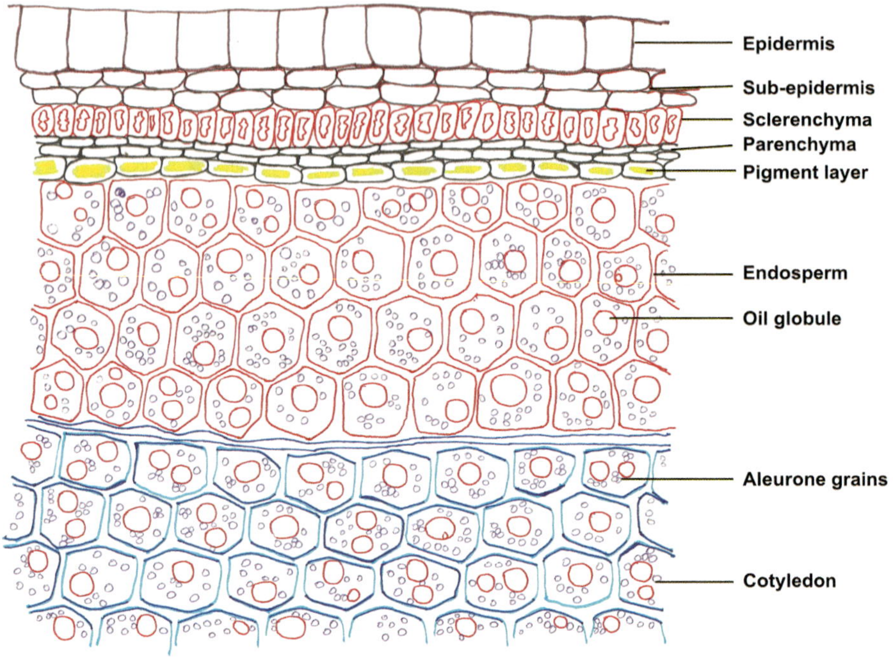

T. S. of Linseed

Powder Character:

Powdered or 'crushed linseed' is a coarse, yellowish-brown powder with dark brown fragments. It possesses a slight odour and an oily mucilaginous taste.

It shows the following characters microscopically:

a. Sclerenchyma:
   The sclerenchymatous layer of testa appears as colourless to pale brown. Cells are lignified and longitudinally elongated with bluntly pointed ends and pitted walls. In some fragments, walls of the cells are very thick and the lumen appears very small like an irregular line. In other cells, walls are thinner and the lumen is narrow.
b. Epidermis of the testa:
   Fragments of the epidermis of the testa are composed of large, thin-walled, polygonal to rounded cells filled with mucilage which get stained with ruthenium red. This is found attached to the parenchymatous layer.
c. Pigment layer of the testa:
   Fragments of the pigment layer of the testa appear abundantly. The cells are square to polygonal with thick, pitted and colourless walls. Each cell is filled with a homogenous mass of orange–brown pigment. These masses are also found scattered and exhibit an important characteristic feature of the powder. Some fragments are found attached to endosperm.

d. Parenchyma of the testa:
   It is composed of one or two layers of cells, rounded to polygonal with thin or thickened walls and irregular intercellular spaces. These layers are found attached to the epidermis and/or to the sclerenchymatous layer.
e. Hyaline layer of the testa:
   It is generally found associated with sclerenchymatous layer. It is composed of very thin-walled, elongated, collapsed indistinct cells. Their long axes are at right angle to those of the sclerenchymatous cells.
f. Endosperm and cotyledons:
   The parenchyma of the endosperm and cotyledons occur abundantly as irregular polygonal cells with moderately thickened walls. These cells contain aleurone grains and globules of fixed oil. Calcium oxalate crystals and starch grains are absent.

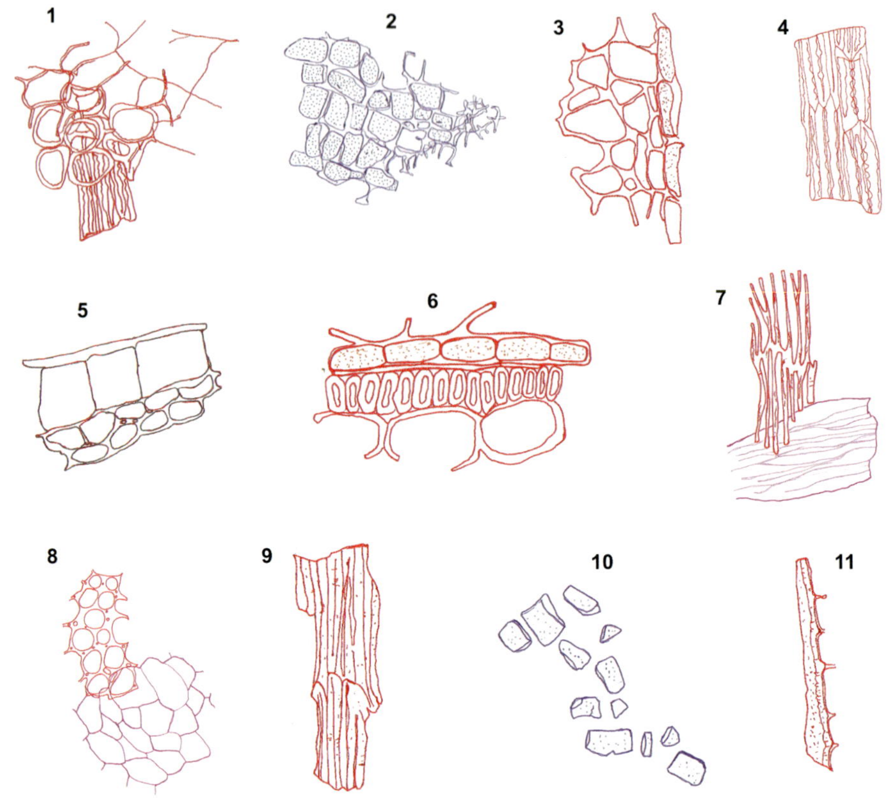

1. **Part of the testa in surface view showing the epidermis, two layers of parenchyma and sclerenchyma**
2. **Pigment layer of the testa in surface view showing the cells containing pigment and underlying endosperm cells**
3. **Pigment layer of the testa and endosperm in sectional view**
4. **Moderately thickened cells of the sclerenchyma layer in surface view**
5. **Epidermis and two layers of parenchyma of the testa in sectional view**
6. **Part of the testa in sectional view showing the sclerenchymatous layer, collapsed hyaline layer and pigment layer**
7. **Thin-walled cells of the sclerenchyma layer in surface view, with associated hyaline layer**
8. **Part of the testa in surface view showing the epidermis and underlying parenchyma**
9. **Thick walled cells of the sclerenchyma layer in surface view**
10. **Isolated masses of pigment**
11. **Part of a cell of the sclerenchymatous layer in longitudinal sectional view showing the dentate outline corresponding to the adjacent parenchymatous cells**

24 Liquorice (*Glycyrrhiza glabra*)

Synonyms:

Glycyrrhiza, Liquorice root, Radix glycyrrhizae

Biological Source:

It consists of dried roots and stolons of the plant *Glycyrrhiza glabra* Linn.

Family:

Leguminosae

Microscopy:

The outline of the transverse section of the liquorice stolon is more or less rounded. The section shows the following characters microscopically:

A. Periderm:
   Phellem, phellogen and phelloderm are collectively known as periderm:

   i. Phellem (cork):
      Transverse section of an unpeeled stolon shows many (about 10–20) radially arranged rows of thin-walled, polygonal, tabular, narrow cork cells. The outer layers contain reddish-brown amorphous matter, and inner layers show thick-walled colourless cells.
   ii. Phellogen (primary cortex):
      It is not distinct, but beneath cork cells there may be a few rows of parenchyma forming the primary cortex.
   iii. Phelloderm (secondary cortex):
      Below the cork is the phelloderm consisting of two to five rows of radially arranged parenchymatous cells whose corners are thickened with cellulose

and some of which may become collenchymatous. Some cells contain simple starch grains and a few contain prisms of calcium oxalate.

B. Secondary phloem:
It is a wide zone which is composed of phloem fibres, phloem parenchyma, medullary rays and cambium:

i. Phloem fibres:
Phloem fibres are thick walled, with lignified outer part and cellulosic inner part arranged as numerous concentric bundles (10–20) of fibres. Each bundle is surrounded by a parenchymatous sheath, each cell of which contains prisms of calcium oxalate. Soft phloem elements, i.e. sieve tissue, alternate with the fibres radially.
ii. Phloem parenchyma:
These phloem cells are thin walled and parenchymatous which contain starch grains and calcium oxalate crystals.
iii. Medullary rays:
Fibres alternate tangentially with the medullary rays. These rays are distinct, bi- to multiseriate and composed of cellulosic parenchyma, with rectangular and somewhat radially elongated cells. These rays appear narrow in the xylem region and wider and funnel shaped in the phloem region. Few cells contain calcium oxalate crystals (prisms) and rounded starch grains.
iv. Cambium:
It appears as an incomplete line composed of about two to three rows of thin-walled, flattened cells.

C. Secondary xylem:
It consists of vessels, wood (xylem) fibres, xylem parenchyma and medullary rays. It is distinct and divided transversely by medullary rays at regular intervals (like the phloem region).

Vessels are thick, yellowish, strongly lignified and large with reticulate, scalariform or pitted thickened walls. There are slitlike bordered pits. These vessels occur singly or in small groups and alternate with bundles of xylem fibres and are surrounded by a sheath of parenchyma containing solitary prism of calcium oxalate in each cell. This is known as 'crystal sheath'.

Xylem fibres are lignified and resemble the phloem fibres. These are also encircled by the crystal sheath.

Xylem parenchymatous cells have lignified pitted walls and contain rounded starch grains. Some cells contain calcium oxalate crystals.

Medullary rays are three to five cell wide in the xylem region and are parenchymatous. Some cells contain starch grains and calcium oxalate crystals.

D. Pith:
It is absent in roots. It consists of large, abundant, thin-walled parenchyma with intercellular spaces. Cells contain starch grains and calcium oxalate crystals. The other tissues appear similar in the root as these are in stolon of liquorice. At the centre there are four small primary xylem bundles arranged at right angles

to each other, with the protoxylem groups directed outwards. These four bundles are separated by an excessive development of parenchyma at the centre of the root. A few rows of phelloderm may be found below the cork.

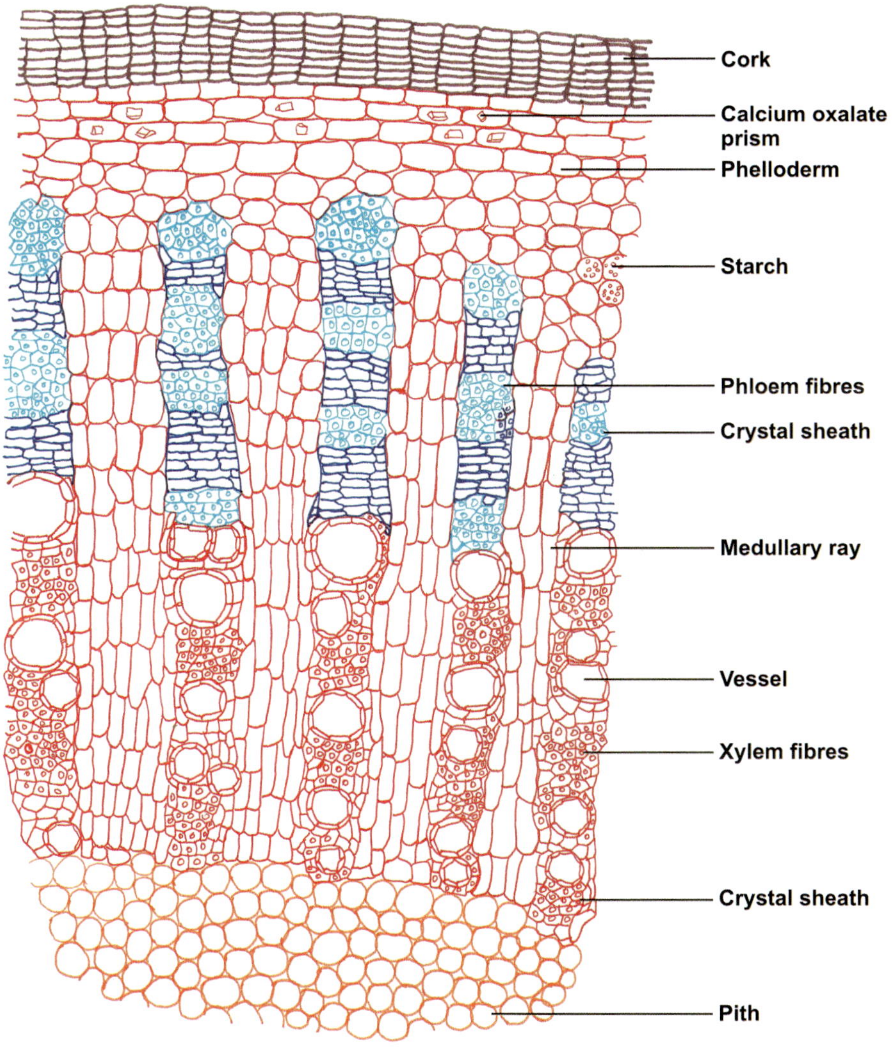

T. S. of Liquorice root

Powder Character:

The powder of liquorice appears pale yellowish brown in colour with faint odour and sweet taste. It shows the following characters microscopically:

a. Fibres:
   These are abundantly seen, as groups of ten to fifteen fibres surrounded by crystal sheath of prisms. Each fibre is lignified, yellow and thick walled, with few small pits. Different layers appear in the walls and therefore the lumen is not clearly visible.
b. Vessels:
   Xylem vessels are found singly or in small groups. Individual vessel is large and found fragmented. These are lignified and have bordered pits. The large vessels are found associated with lignified xylem parenchyma composed of moderately thin and pitted walled, square to elongated cells, rectangular in outline.
c. Starch grains:
   Most of the starch grains are simple and small, oval to elongated, rounded, spherical and slightly flattened. Larger grains may show a slit-shaped hilum. A few compound grains may be observed with two to four components.
d. Calcium oxalate crystals:
   Prisms of calcium oxalate which are uniform in size are observed within the cells which form the 'crystal sheath' surrounding the fibres. A few larger prisms or twin prisms also occur in some of the parenchymatous cells of the medullary rays and pith. Some are found as scattered in the powder.
e. Cork cells:
   Fragments of cork cells are seen as orange brown, thin walled and polygonal in surface view. These are abundantly found in the case of unpeeled liquorice.
f. Parenchyma:
   Thin-walled parenchymatous cells occur abundantly which are from the cortex, medullary rays, pith and xylem parenchyma. These cells are rounded to rectangular in outline and are filled with starch granules. Sometimes phloem parenchyma or xylem parenchyma cells are associated with medullary rays. A small amount of collenchyma is also present.

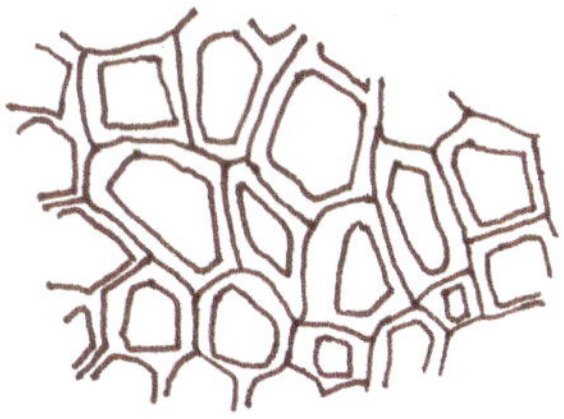

Cork cells
in surface view

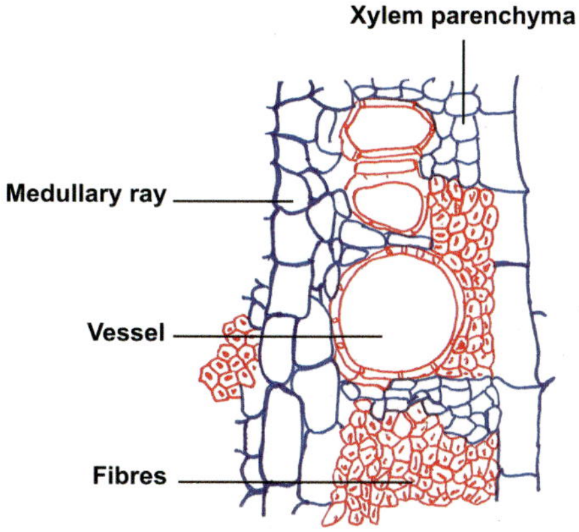

T. S. through xylem portion

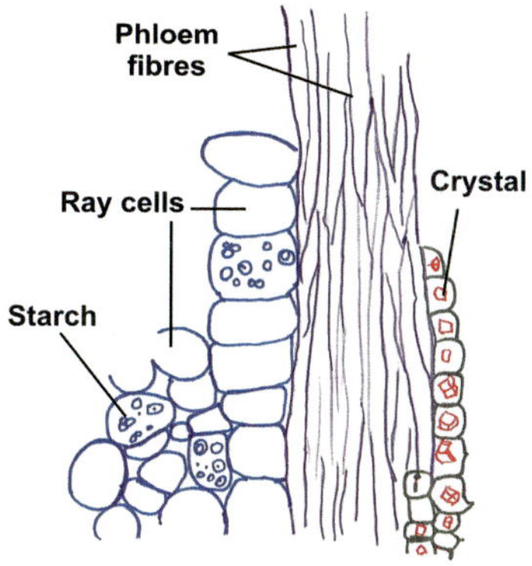

L. S. through phloem region

Starch granules

Calcium oxalate crystals

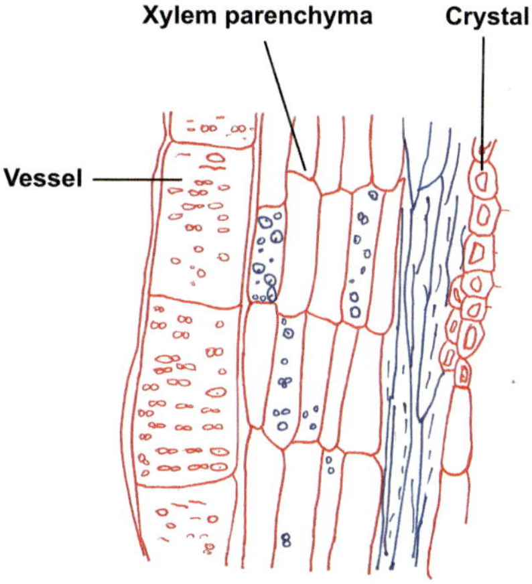

L. S. through xylem portion

25 Neem (*Azadirachta indica*)

Synonyms:

Margosa, Nimba, Limba, Neem, Nimb

Biological Source:

It consists of dried leaflets of the plant *Azadirachta indica* A. Juss.

Family:

Meliaceae

Microscopy:

Microscopically neem leaf shows a typical dorsiventral pattern. The following characters are observed in the transverse section in the lamina and midrib region:

A. Lamina:
   It is composed of the upper epidermis, palisade, spongy parenchyma and lower epidermis:

   i. Upper epidermis:
      It is made up of polygonal cells arranged in a single row with no stomata. The outer walls of cells show a thick cuticle.
   ii. Palisade:
      Two layers of palisade cells are observed below the upper epidermis. Occasionally a few of these cells contain rosette crystals of calcium oxalate.

iii. Spongy parenchyma:
The spongy parenchyma is characterised by the presence of intercellular spaces and vascular bundles. The bundles indicating the positions of veins are interspaced within spongy parenchyma. Secretary cells are abundantly seen mainly on the borderline of spongy parenchyma and palisade. A few cluster crystals of calcium oxalate are also observed.

iv. Lower epidermis:
It resembles the upper epidermis in many respects. Cells are smaller than the upper epidermis. Anomocytic stomata are numerous on the lower epidermis.

Leaf bears covering trichomes on both epidermises which are unicellular and curved (occasionally bicellular and uniseriate) with pointed apex.

B. Midrib:
The midrib region shows ventral and dorsal ridges, composed of collenchymas. The cortical region of the midrib shows rosette crystals of calcium oxalate. The arc-shaped xylem vessels with spiral/pitted thickening are also observed along with phloem. The transverse section of the rachis shows a single layer of epidermis, six to eight layers of cells forming the cortex. Numerous secretary cells are found as scattered.

Groups of fibres are scattered in the phloem region. Xylem and pith are made up of cells with intercellular spaces.

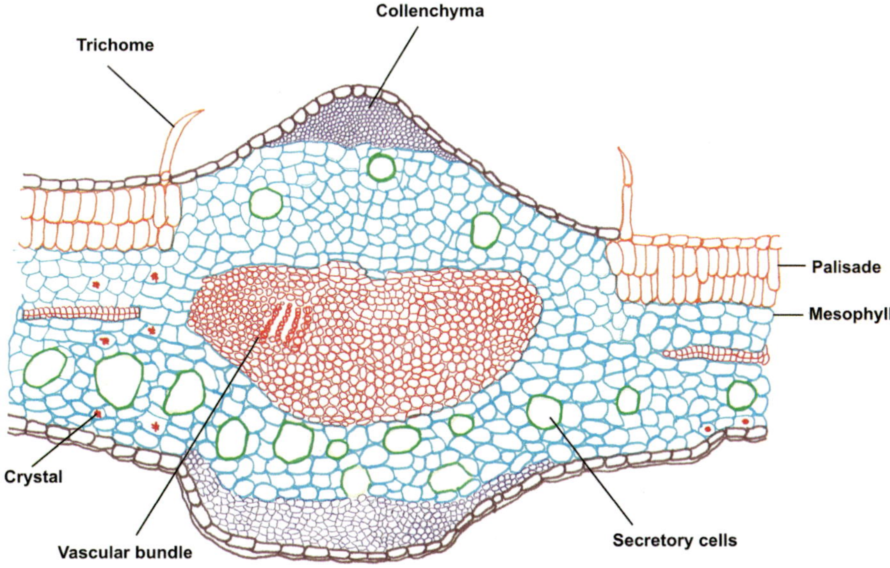

T. S. of midrib of the Neem leaf

Powder Character:

The powder of dried leaves of neem is greenish yellow to faint brown with characteristic faint odour and intensely bitter taste. It shows the following characters microscopically:

a. Fragments of epidermis:
   Cells of the upper epidermis appear as polygonal and without any stomata. Lower epidermal cells show presence of anomocytic stomata.
b. Fragments of palisade:
   Palisade cells are observed as fragments as well as whole cells arranged in a double layer. Secretary cells also appear occasionally along with palisade cells.
c. Vascular tissues:
   Vascular tissues are observed along with adjacent cells.
d. Cortical cells:
   Cortical region of the rachis shows rosette crystals of calcium oxalate in the parenchyma.
e. Trichomes:
   These are observed on both epidermises as covering, uniseriate and unicellular (sometimes bicellular) with sharp apex.
f. Vessels:
   Spiral and pitted vessels are observed in the powder.
g. Pith cells:
   These are large and have intercellular spaces.
h. Spongy parenchyma:
   Cells of spongy parenchyma show secretory cells, vascular tissues and intercellular spaces.

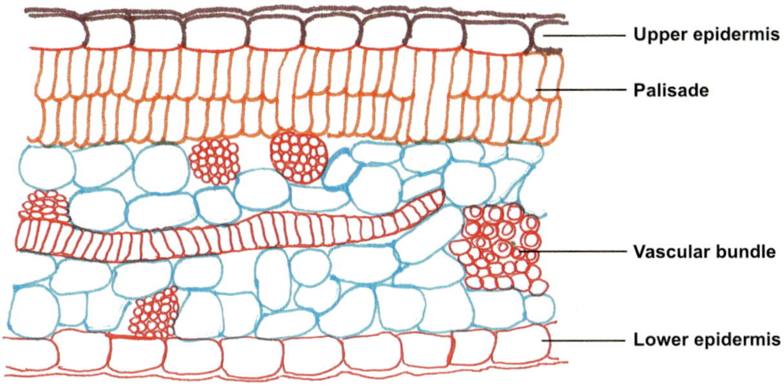

T. S. of the leaflet through lamina

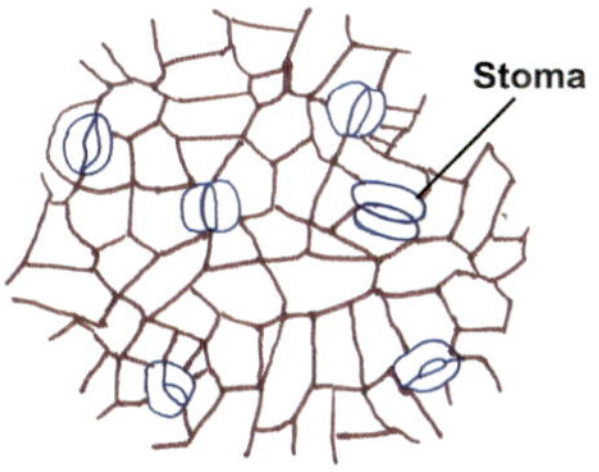

Lower epidermis in surface view

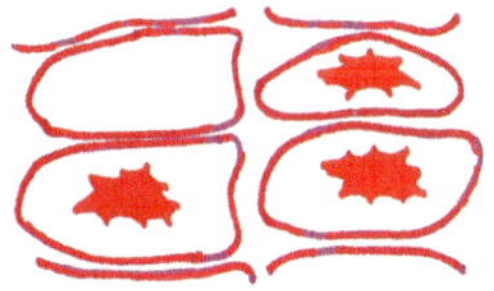

Parenchyma from the cortical region of rachis with rosette crystals

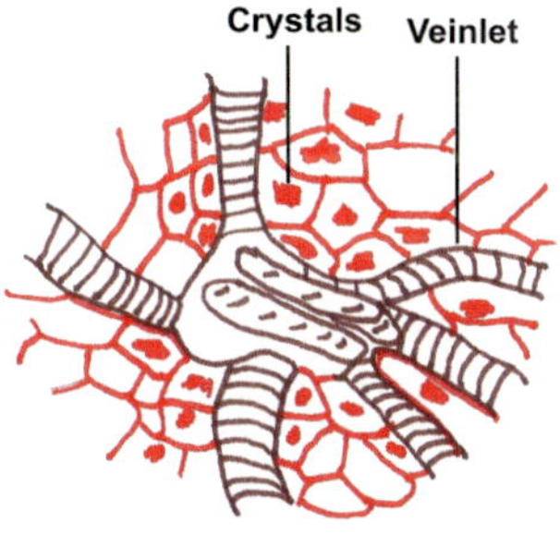

Veinlet with crystals

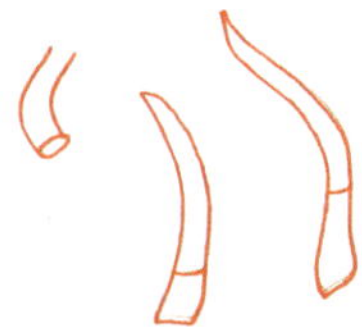

Covering trichomes

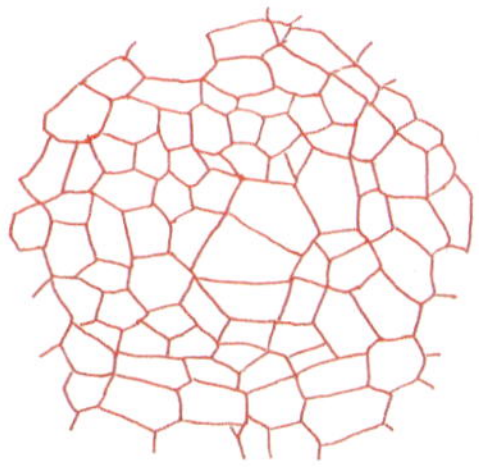

Upper epidermis in surface view

26 Nutmeg and Mace (*Myristica fragrans*)

Synonyms:

Myristica, *Nux moschata*, Nutmeg

Biological Source:

It consists of dried kernels of the seeds of the plant *Myristica fragrans* Houtt.

Family:

Myristicaceae

Microscopy:

In order to perform the systematic study under the microscope, longitudinal and transverse sections must be taken for nutmeg along with the powder of the kernel. Longitudinal section is lustrous and has a marbled look.

The outer tissue is the perisperm, which possesses fibrovascular bundles. These bundles and their positions can be confirmed by the reticulate furrows on the surface.

A. Perisperm:
   Perisperm along with light brown endosperm penetrates the branching and thus possesses marbled appearance:

   i. Outer perisperm:
      The cells of the outer perisperm are radially flattened. These cells contain brownish fat, which is insoluble in potassium hydroxide as well as in chloral hydrate solution. Sometimes prismatic or disc-shaped crystals are observed which may be of potassium acid tartrate.

ii. Inner perisperm:
The cells of inner perisperm show many lamellae along the furrows on the surface and seem as penetrating into the endosperm. These lamellae are made up of small parenchymatous cells with thin walls and brownish contents as well as rounded, oval oil cells which occur singly or in groups. Small groups of lignified spiral vessels associated with inner perisperm are also observed.

B. Endosperm:
It is made up of parenchymatous cells with thin brown cell walls, closed, packed, polygonal and filled with starch granules. Starch granules are simple or compound (with 2–10 components). Some of the cells contain aleurone grains with a crystalloid and feathery fat crystals. Very few tannin cells containing tannin and starch occur as scattered in the endosperm.

Powder Character:

The powder of nutmeg is cinnamon brown in colour and possesses characteristic, aromatic odour and an aromatic slightly bitter taste.

It shows the following specific features under the microscope:

a. Starch grains:
Abundant starch granules, few simple and spherical but mostly of compound nature with two to eight (up to ten) components are seen. These are large and show central stellate/slit-shaped hilum.
b. Cells of perisperm:
The cells of the outer perisperm parenchyma are pale and composed of polygonal to rounded cells with slightly thick walls and small intercellular spaces. A few of these cells contain prisms.
c. Parenchyma:
The parenchyma of inner and ruminating perisperm is composed of smaller cells with reddish-brown contents and large rounded oil cells. Groups of lignified spiral vessels are occasionally found.
d. Endosperm:
Parenchyma of the endosperm is made up of thin-walled, closely packed polygonal cells with starch granules. Few of the cells contain small elongated prisms.
e. Cell contents:
Crystals of the fat form large feathery or irregularly shaped masses when solution of chloral hydrate and powder are mounted, heated and allowed to cool.

27 Mace

Common mace consists of dried arillus or arillode of *Myristica fragrans* seed. It occurs as yellowish-orange strips or coarsely reticulate bands or as a yellowish- to orange–brown gritty powder. The colour of the fresh mace is bright red. Odour and taste are aromatic and similar to nutmeg. The mace powder consists of mainly parenchymatous ground tissue along with large yellowish-brown oil cells and occasional crystals. Small groups of lignified vessels are also present. Abundant fat is observed which forms large feathery masses in a cooled chloral hydrate mount. The powder shows presence of fragments of epidermis made up of narrow, elongated cells with thin slightly sinuous walls. It also contains small irregular granules of amylodextrin which gives red colour with iodine solution.

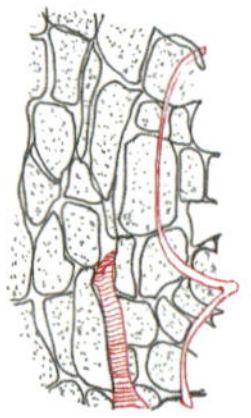

Inner or ruminating perisperm with parts of oil cells and a fragment of vascular tissue

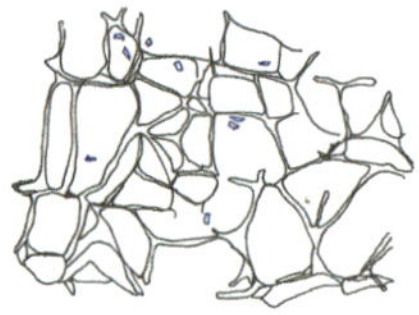

Outer layers of the perisperm with scattered prisms

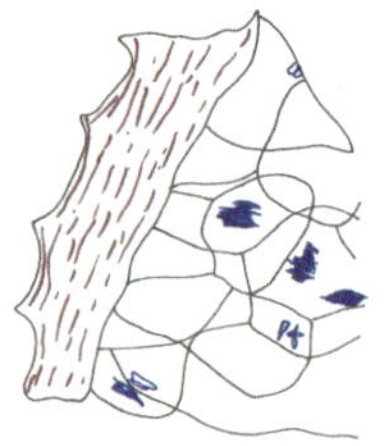

Cells of the outer part of the endosperm containing crystals, with part of the adjacent ruminating perisperm

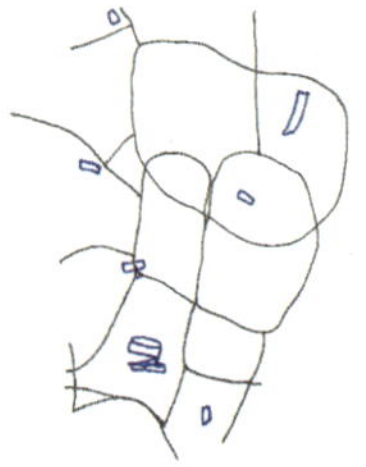

Cells of the inner part of the endosperm, some containing prisms

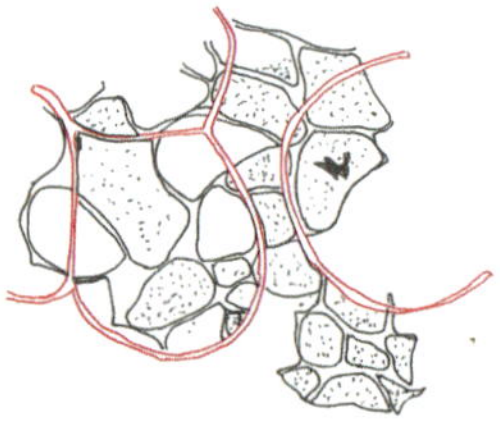

Inner or ruminating perisperm with oil cells

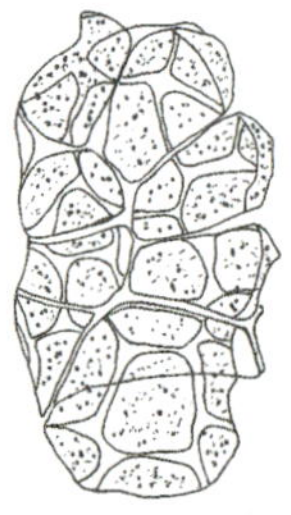

Layers of the perisperm containing pigment

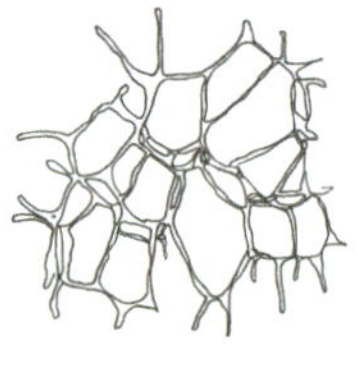

Outer layers of the perisperm in surface view

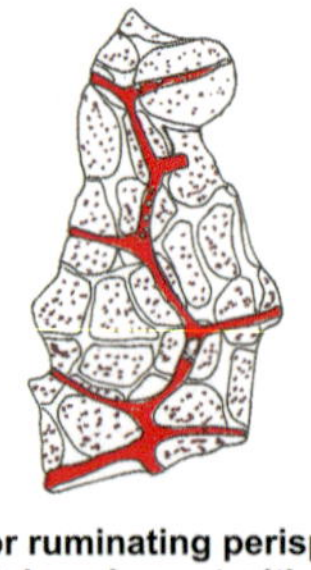

Inner or ruminating perisperm containing pigment with part of a group of oil cells

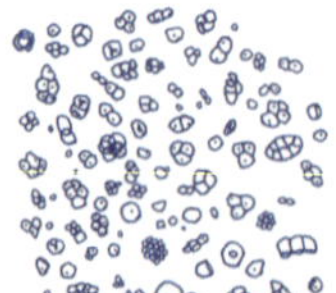

Starch granules

Crystalline masses of fat from a cooled chloral hydrate mount

28 Nux vomica (*Strychnos nux-vomica*)

Synonyms:

Nux vomica, Crow fig, Semina strychni

Biological Source:

It consists of all dried ripe seeds of the plant *Strychnos nux-vomica* Linn.

Family:

Loganiaceae

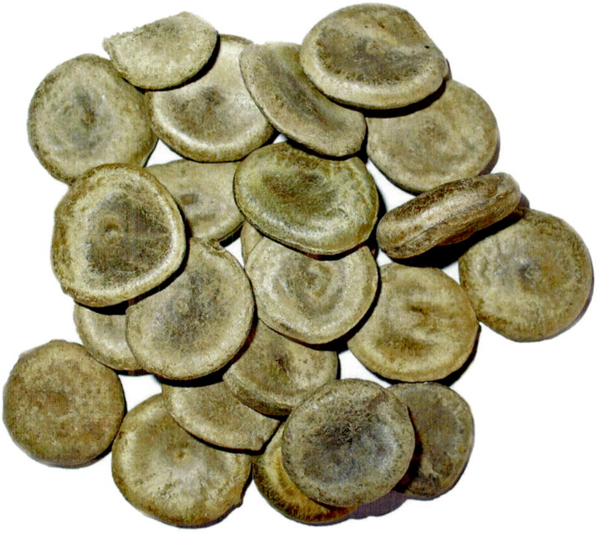

Microscopy:

Transverse section of the nux vomica seed shows a hairy testa and a bulky endosperm. The following tissues are observed microscopically:

A. Testa:
   The testa is thick and the major portion is occupied by the epidermis:

   i. Epidermis:
      Each epidermal cell is extended to form an appressed trichome. These trichomes are characteristic and lignified. These have large and broad basal portion and are wide. The upper portions of trichomes have about ten longitudinal ridges, united by a thin wall, and are placed almost at right angle to the bases and radiate out towards the margin of the seed. Sometimes trichomes appear as bent, twisted and parallel in one direction. Surface irregularities in the basal region of trichomes cause them to interlock

with one another. This arrangement of trichomes gives a silky appearance to the surface (testa) of the seed. The epidermis is composed of a single layer of thick-walled cells. Many times walls show small pits, i.e. bases of trichomes.

ii. Collapsed parenchyma:
A double layer of flattened parenchymatous cells is seen below the epidermis.

B. Endosperm:
It occupies the major portion of the seed. It is composed of thick- walled non-lignified cellulosic isodiametric parenchymatous cells. Outer layers of cells (below the collapsed parenchyma) appear palisade like, but cells towards the inner side are larger. The cell walls are mainly composed of hemicelluloses and swell well in water. The lumen of cell is polygonal.

i. Plasmodesmata:
An important characteristic feature of the endosperm of the nux vomica seed is the presence of well-developed 'plasmodesmata,' i.e. very fine protoplasmic strands between the walls of endospermic cells. The protoplasts of the endospermic cells communicate through the cell walls with plasmodesmata. These appear as very fine lines crossing the walls and can be seen more clearly by staining the section with dilute iodine solution.

ii. Aleurone grains:
These grains contain fixed oil and are of irregular shapes. Several globoids are present in each grain.

iii. Oil globules:
Fixed oil as small droplets can be observed in the cells. Endospermic cells contain strychnine in the inner part and brucine in the outer layers of the endosperm. Solution of ammonium vanadate in sulphuric acid gives a violet colour to the section revealing presence of strychnine. A crimson colour is produced when the section is mounted in nitric acid because of brucine.

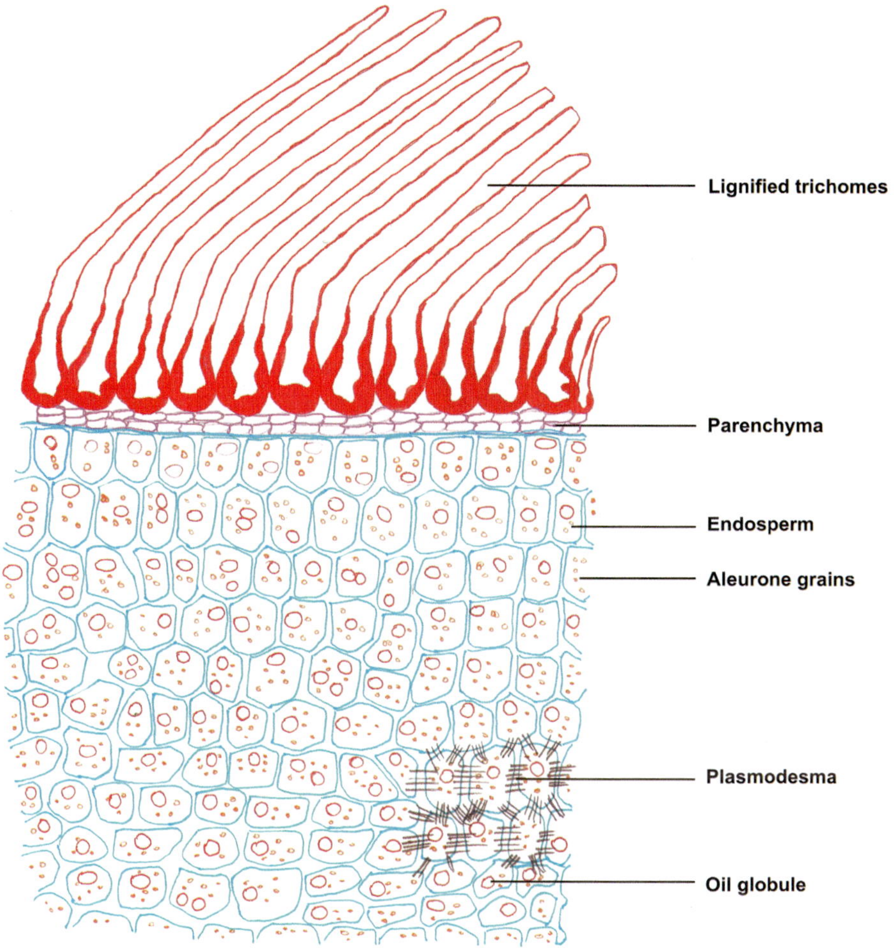

T. S. of Nux-vomica seed

Powder Character:

The powder of nux vomica seeds is yellowish grey to brownish grey in colour with slight fatty and rancid odour and bitter and persistent taste.

Powder microscopically shows the following features:

a. Endospermic cells:
   Cells of endosperm are seen as fragments or whole. The cells from outer region are small, relatively thin walled, polygonal and slightly elongated. These fragments of cells are found associated with the pigment layer of testa and composed of a layer of indistinct cells containing orange to brown pigment. Endosperm cells from the central region are large and thick walled and have a small lumen. Sometimes cells show presence of plasmodesmata in the walls. Endosperm also shows aleurone grains and oil globules within the cells.
b. Trichomes:
   Fragments or whole trichomes are seen. These appear as narrow, lignified, aggregated rods running longitudinally. These are cylindrical and of varying lengths and thicknesses. Trichomes have broad base and rounded apex and are pitted. A few small ridges are also observed on the surface of the trichomes.
c. Testa:
   The epidermis of testa is sclerenchymatous and seen as a single row of yellowish-brown cells. These cells are extended to build a trichome and are thick walled and pitted. About ten lignified rods form each trichome longitudinally. Trichomes are broken off and broken ends of these lignified rods remain attached to the epidermal cells.

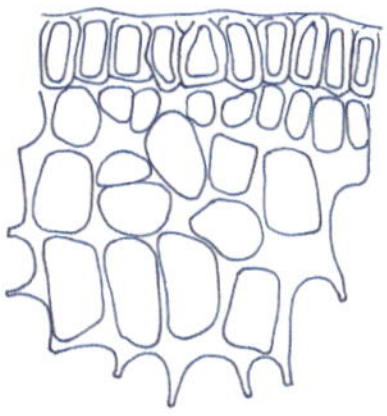

Outer part of the endosperm in sectional view

An almost complete trichome showing the rounded apex

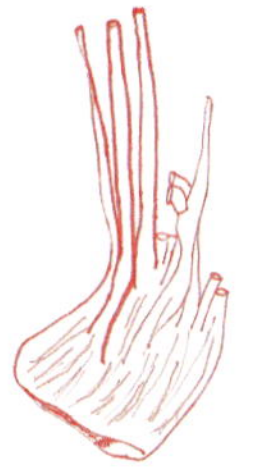

Part of the sclerenchymatous epidermis of the testa in side view, with parts of the trichome attached

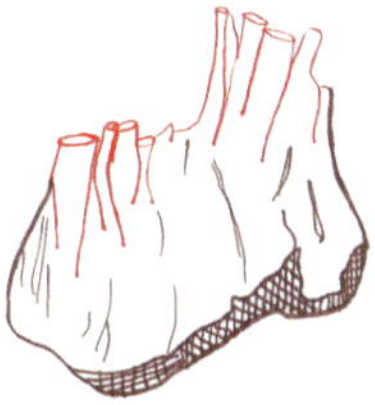

Part of the sclerenchymatous epidermis of the testa in side view, with broken-off remains of the trichome rods attached

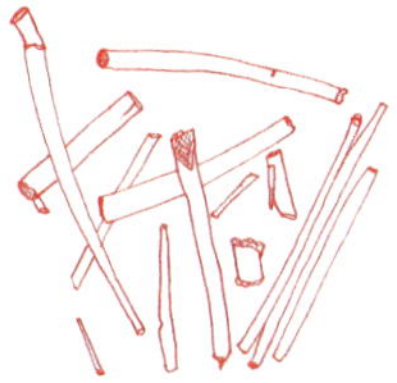

Fragments of trichome rods

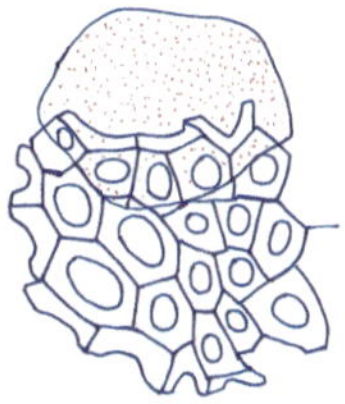

Outer layer of the endosperm in surface view, with associated pigment layer of the testa

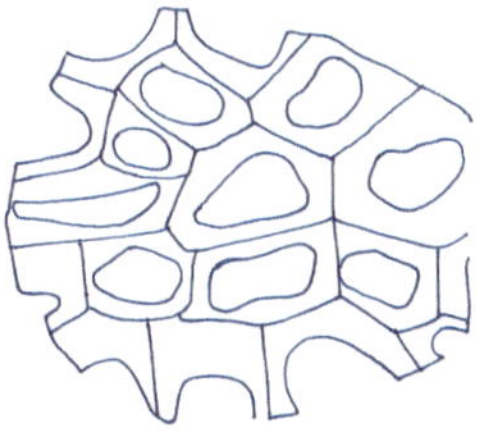

Endosperm cells from the central region

Part of a large trichome

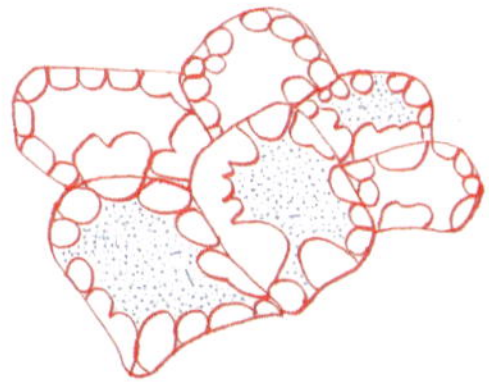

Sclerenchymatous layer of the testa in surface view showing the bases of the lignified rods and pigment in some of the cells

29 Opium (*Papaver somniferum*)

Synonyms:

Raw opium, Opium, Opium poppy

Biological Source:

It is the milky exudation (latex) obtained by making incisions into the unripe capsules of the plant opium poppy, *Papaver somniferum* var. *album* Linn., and dried naturally by spontaneous evaporation and partly by artificial heat.

Family:

Papaveraceae

Microscopy:

Opium consists of the dried latex obtained from the incised poppy capsules. Powdered opium is brownish in colour with characteristic odour and bitter taste. It consists of numerous brown, agglomerated, granular, irregular masses of dried latex along with small particles of vegetable tissues. The latex is water soluble and has no cellular structure. The residue left after water extraction of opium is a mixture of other small amounts of characteristic material. These particles of other vegetable tissue material are seen as a result of a method of collection and preparation process. These particles mainly include commonly observed fragments of outer capsule wall and pollen grains. If the sample is contaminated with other parts of capsule, the inner epidermis of the capsule, epidermis of the placenta and fragments of vessels are seen. Some varieties of opium are covered with the coarse powder of poppy leaves, and thus upper and lower epidermises of the leaves are observed occasionally.

A standard powder containing a definite percentage of morphine is adjusted, if needed by adding the calculated amount of powdered cocoa shells (husk). Therefore, the opium may show a layer of small, polygonal cells of cocoa husk under the microscope.

These vegetable tissues are described below:

a. Fragments of outer epidermis of the capsule:
   These are made up of unlignified polygonal tabular cells with unevenly thick anticlinal walls and show rounded anomocytic stomata. Sometimes these walls are pitted. Cells of the inner layers are collenchymatous.
b. Pieces of upper and lower epidermises of the foliage leaves of *Papaver somniferum* are composed of thin-walled polygonal cells with many ranunculaceous (anomocytic) stomata on the lower epidermis. Palisade cells beneath the upper epidermis are large and loosely packed. Spongy mesophyll is seen rarely as attached to the lower epidermis.
c. Pollen grains:
   These are seen occasionally as spherical, smooth with three pores and faintly pitted exine.
d. Cells of inner epidermis of the capsule wall:
   These should be absent, but if present these occur as longitudinally elongated, with lignified and thickened walls. The anticlinal walls appear pitted and show a few pits on the inner walls. In surface view, cells appear as polygonal and thick walled with a stellate lumen. Large undeveloped stomata occur scattered and give negative reaction for lignin.
e. Starch:
   Small traces of starch from the capsule walls may occur; grains are rounded.
f. Epidermis of the placenta:
   It is made up of cells with thick and lignified walls with slit-shaped pits. In the surface view the walls appear beaded.
g. The characteristic layer of small, polygonal cells of cocoa husk.
   These have moderately thick walls. This layer is usually known as the sclerenchymatous layer, but it does not give reaction with solution of phloroglucinol and hydrochloric acid. In surface view, sometimes small groups of cells are seen formed as a result of splitting of cells.
h. Spongy parenchyma of cocoa husk.
   Fragments of spongy parenchyma of the cocoa husk consist of rounded cells with slightly thickened walls and irregular intercellular spaces. Lignified, spirally thickened vessels occur as embedded in the parenchyma. These vessels are fairly uniform in diameter and occur singly or in small groups.

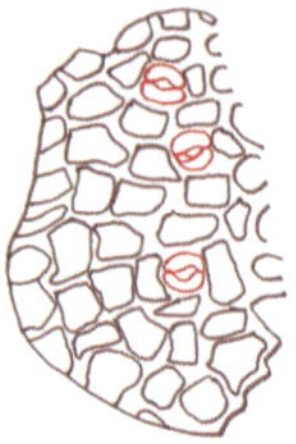

Outer epidermis of the capsule in surface view

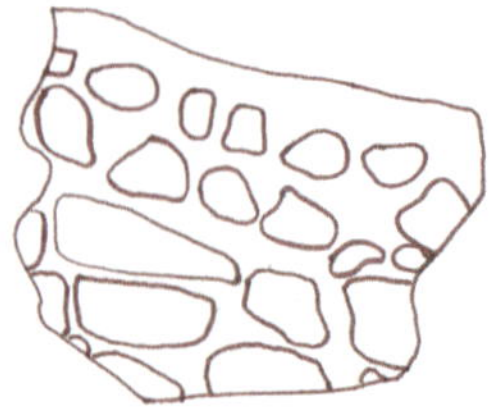

Outer layers of the capsule in sectional view

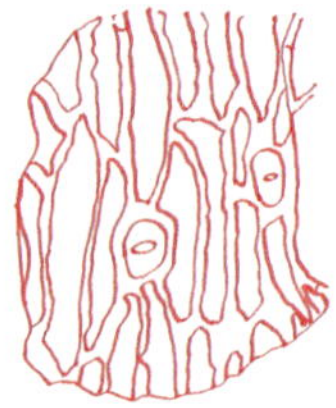

Inner epidermis of the capsule in surface view

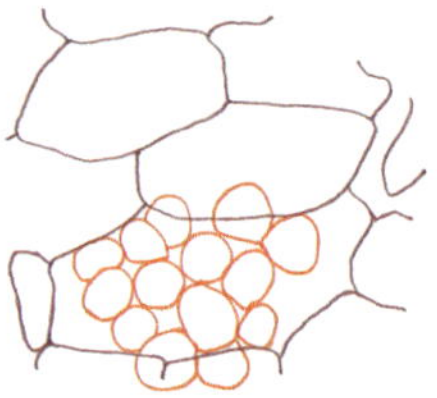

Upper epidermis of the leaf in surface view with part of the underlying palisade

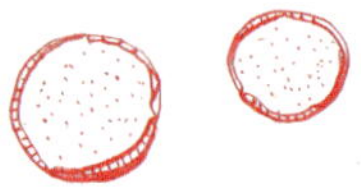

Pollen grains

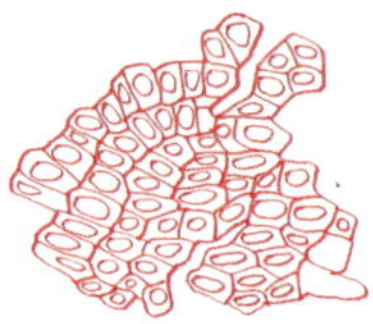

Parts of the layer of thick-walled cells of cocoa husk in surface view

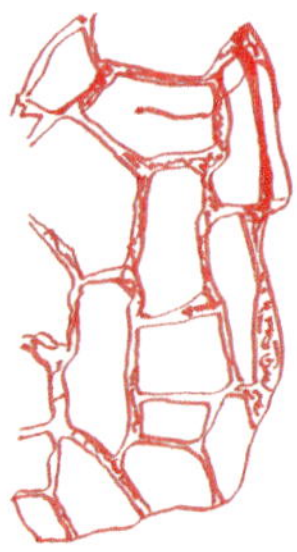

Epidermis of the placenta in oblique surface view

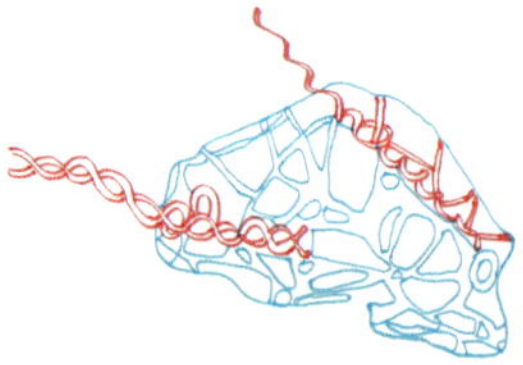

Spongy parenchyma and spiral thickenings from the middle layers of cocoa husk

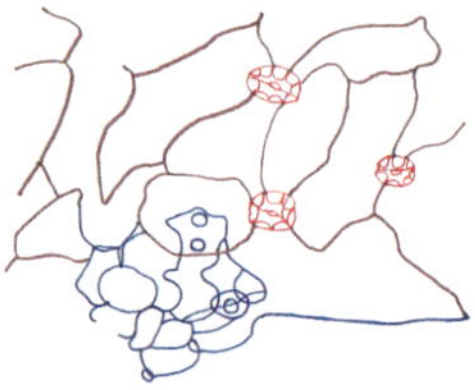

Lower epidermis of the leaf in surface view showing anomocytic stomata and part of the underlying spongy mesophyll

30 Podophyllum (*Podophyllum emodi*)

Synonyms:

Podophyllum radix, Indian podophyllum

Biological Source:

It consists of dried pieces of rhizomes and roots of plants *Podophyllum hexandrum* Royle or *Podophyllum emodi* Wel.

Family:

Berberidaceae

Microscopy:

The transverse section of the rhizome of *Podophyllum* microscopically shows the following tissues from the periphery towards the centre:

A. Epidermis:
   It is dead sometimes. It consists of elongated, rectangular, tabular, dark reddish-brown cells. A few cells are isodiametric.
B. Cork:
   It is made up of about one to six layers of tabular, thin-walled, polygonal cells with brownish content.

C. Cortex:
   It is abundant and forms the major bulk of the ground tissue. It consists of pitted, stout-walled cellulosic parenchyma with a small portion of collenchyma in the outer region. Most of these cells contain starch grains. These starch grains are usually simple but sometimes compound made up of two, four, eight or more (up to 15–20) components. Some cells of the ground tissue show presence of cluster crystals of the calcium oxalate. The size of the calcium oxalate crystals and starch grains can be a distinguishing feature between Indian podophyllum and American podophyllum. The cortex also shows a few scattered, lignified fibres.
D. Vascular bundles:
   Vascular bundles are small, conjoint, collateral and open with lignified, pitted inner xylem vessels and outer phloem tissue. Above the phloem, lignified pericyclic fibres are present. Vascular tissues are separated by medullary rays. Medullary rays are seen as wide, prominent, multiseriate and made up of thick-walled parenchymatous cells.
E. Pith:
   It is wide and made up of parenchymatous cells. Cells of pith show presence of cluster crystals of calcium oxalate and starch along with brownish content in few cells. A few sub-cylindrical, narrow stone cells occur in the pith along the inner side of some of the vascular bundles.

The transverse section of the podophyllum root shows a circular outline with typical features of a dicot root. The following tissues are seen from the periphery towards the centre:

A. Epiblema:
   It is single layered and slightly papillose with thick yellowish-brown outer walls (cuticle). Outer and radial walls are thick and suberised.
B. Exodermis:
   It is just below the epiblema, single layered and made up of small, thin, wavy-walled, suberised cells.
C. Cortex:
   It is a wide zone of about 15–20 rows of rounded, cellulosic parenchymatous cells. Outer cells are collenchymatous. The cell walls are wavy and pitted and show intercellular spaces. Cells of the cortex contain starch abundantly, and calcium oxalate crystals are absent.
D. Endodermis:
   It is single layer of thick-walled, elongated, suberised cells with distinct, lignified prominent 'Casparian' strips.
E. Vascular bundles:
   These are radial, alternate groups made up of 4–10 bundles and with exarch protoxylem. Vascular bundles are separated by medullary rays.
F. Pith:
   It is mainly composed of a group of lignified, pitted sclereids present at the centre. The xylem groups of roots which are close to the rhizome form a continuous ring.

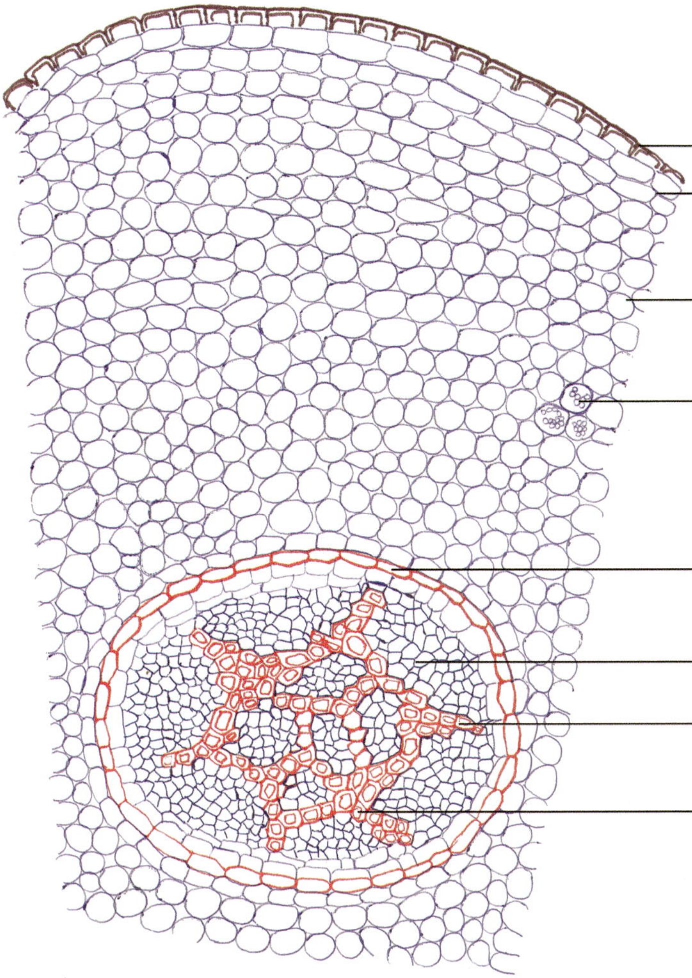

T. S. of Podophyllum root

Powder Character:

The powder of podophyllum is light brown in colour with a slight odour and bitter taste. It shows the following characters microscopically:

a. Sclereids:
   These are abundant and occur in groups. Cells are elongated and rectangular in outline; walls are thickened and pitted. Cells are lignified and groups of sclereids are associated with thin-walled parenchyma of the pith.
b. Cork:
   Fragments appear brownish and are composed of thin-walled, lignified, polygonal cells.

c. Vessels:
   These vessels occur singly or in groups as associated with thin-walled xylem parenchyma. These are lignified, reticulately thickened and irregular in shape. Sometimes vessels show slit-shaped pits and spiral or annular thickening.
d. Starch granules:
   These are abundant, simple, small and spherical to ovoid in shape or mainly compound with two, three or up to eight or up to twenty components.
e. Calcium oxalate crystals:
   Cluster crystals of calcium oxalate are seen. These are large, few in number and found scattered or within the thin-walled parenchyma of the cortex.
f. Epiblema and exodermis:
   Fragments appear brown in colour; cells are elongated to rectangular in outline with thick walls and few pits. Epiblema cells or fragments are associated with exodermis, cells of which are similar in shape and size but with thin and wavy cell walls.
g. Parenchyma:
   These cells are abundantly seen as fully loaded with starch granules. Sometimes cluster crystals of calcium oxalate are seen inside the cells. These cells are thin walled, elongated and rounded and occur in small groups. The parenchymatous cells of the pith have thickened and pitted walls.

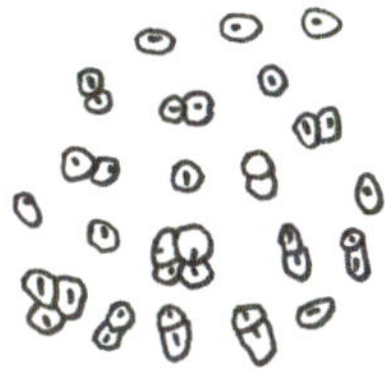

Starch Granules

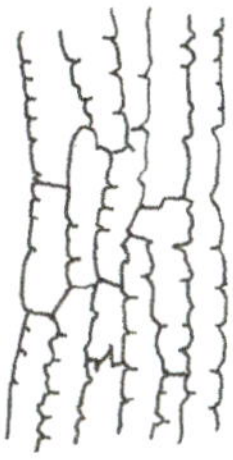

Epidermis in surface view

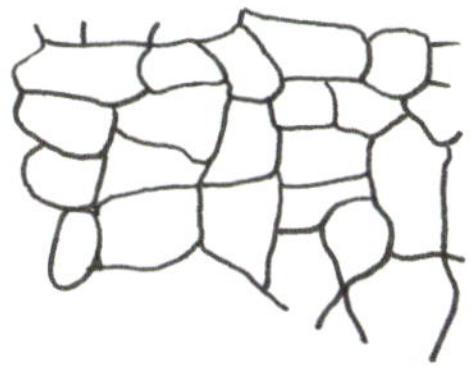

Cork in surface view

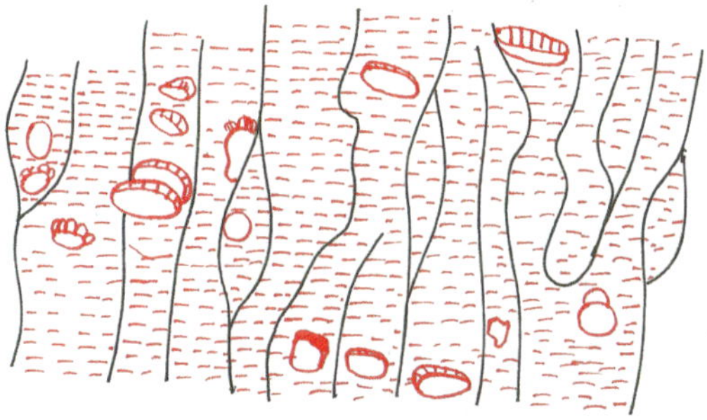

Part of a group of vessels from rhizome

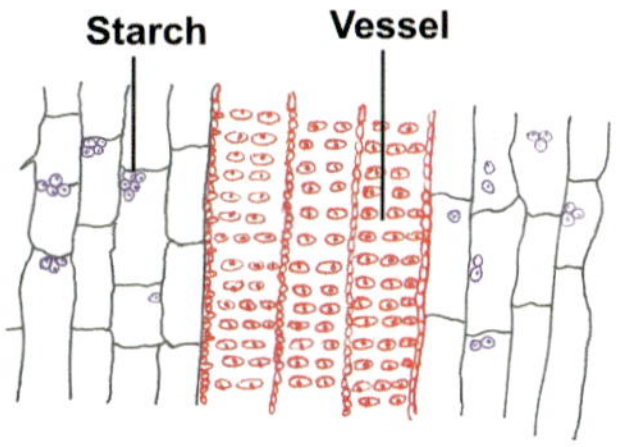

L. S. of root through xylem portion

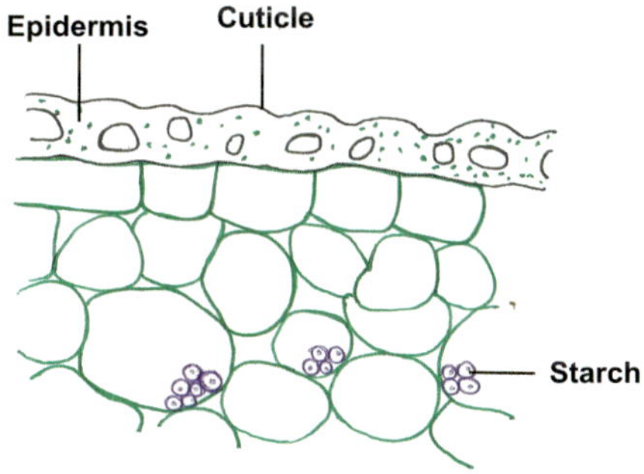

T. S. of part of root showing epidermis and cortex

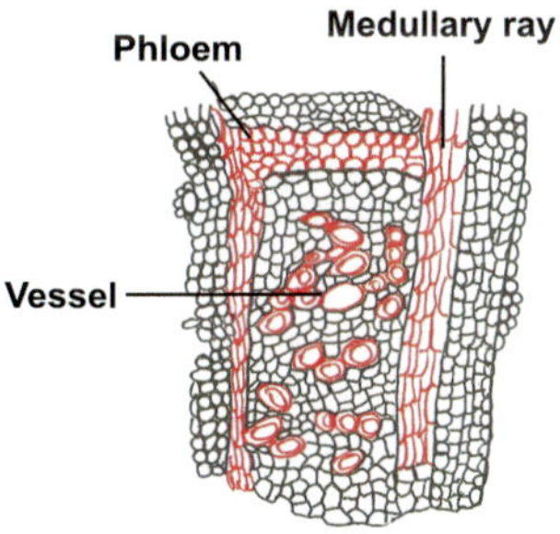

Vascular bundle of rhizome in section view

31 Quassia (*Picrasma excelsa*)

Synonyms:

Quassia wood, Jamaica quassia, Bitterwood, Quassia lignum

Biological Source:

It consists of dried stem wood of the plant *Picrasma excelsa* Planchon.

Family:

Simaroubaceae

Microscopy:

The quassia wood shows a storeyed arrangement of lignified tissues; thus, there different microscopic sections are taken for the microscopic anatomical study, i.e. transverse section (TS), tangential longitudinal section (TLS) and radial longitudinal section (RLS).

A. Transverse section (TS):
The TS shows the following characters:

i. Secondary xylem:
It is composed of xylem vessels, xylem fibres and xylem parenchyma:

a. Xylem vessels:
These are seen as big, numerous and either singly or in groups of about two to ten. These are closely arranged and fill the entire region between the two consecutive medullary rays. These show minute bordered pits, with elliptical or hexagonal borders and slitlike pores. The length of an individual vessel is about one to five times more than that of breadth.

b. Xylem fibres:
These occupy the major portion of the wood. These are arranged in radial rows. In the TS, the fibre mass appears to consist of alternate rows of smaller and larger elements. Fibres have narrow, oblique and slitlike pits.

c. Xylem parenchyma:
The cells of xylem parenchyma are small and square to polygonal or rectangular with uniformly and moderately thickened pitted walls. These cells are arranged in three- to seven-cell thick layers. Some cells of xylem parenchyma contain prisms of calcium oxalate. Each crystal is surrounded by a closely fitting lignified envelope. A few cells show presence of starch grains within them. These grains are simple, spherical or compound with two components.

ii. Medullary rays:
These are prominent and homogenous. These are one to four cells wide and slightly elongated radially. These are heavily pitted with oblique septa. These occasionally contain prisms of calcium oxalate and starch granules.

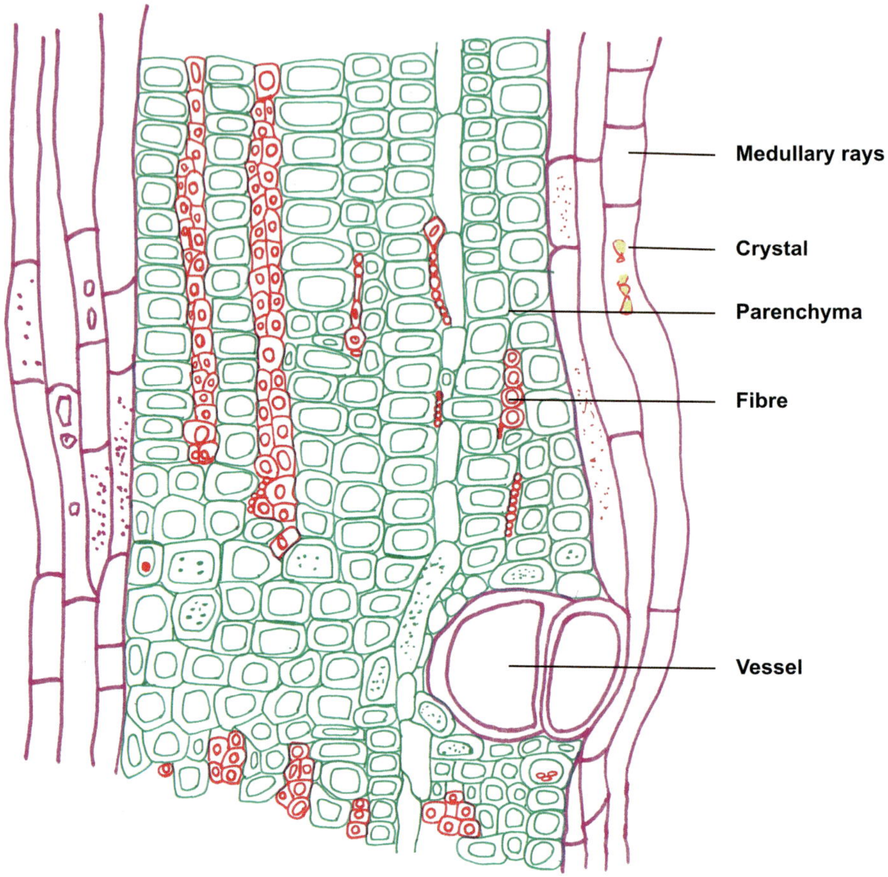

T. S. of Quassia wood

B. Tangential longitudinal section (TLS):
It shows the following features:

i. Secondary xylem:
It consists of xylem vessels, xylem (wood) and xylem parenchyma:

a. Xylem vessels:
These are seen as wide, elongated and with minute bordered pits.

b. Wood fibres:
These are present in groups, long and tapering with small bordered pits.

c. Xylem parenchyma:
It is composed of elongated parenchymatous cells which have pitted walls. Some of these cells show presence of small prisms of calcium oxalate.

ii. Medullary rays:
These are cut transversely and thus height and width can be measured in the TLS.

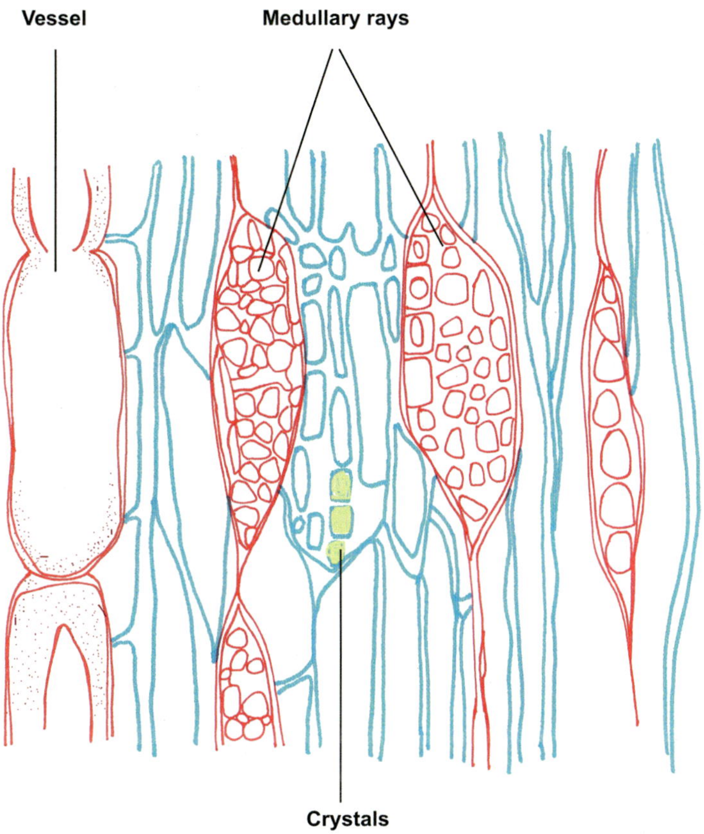

T. L. S. of Quassia wood

C. Radial longitudinal section (RLS):
   It shows secondary xylem and medullary rays:

   i. Secondary xylem:
      It appears same as in TLS.
   ii. Medullary rays:
      These are observed in groups. These rays cut the xylem fibres and vessels at right angles. The length of medullary rays can be measured.

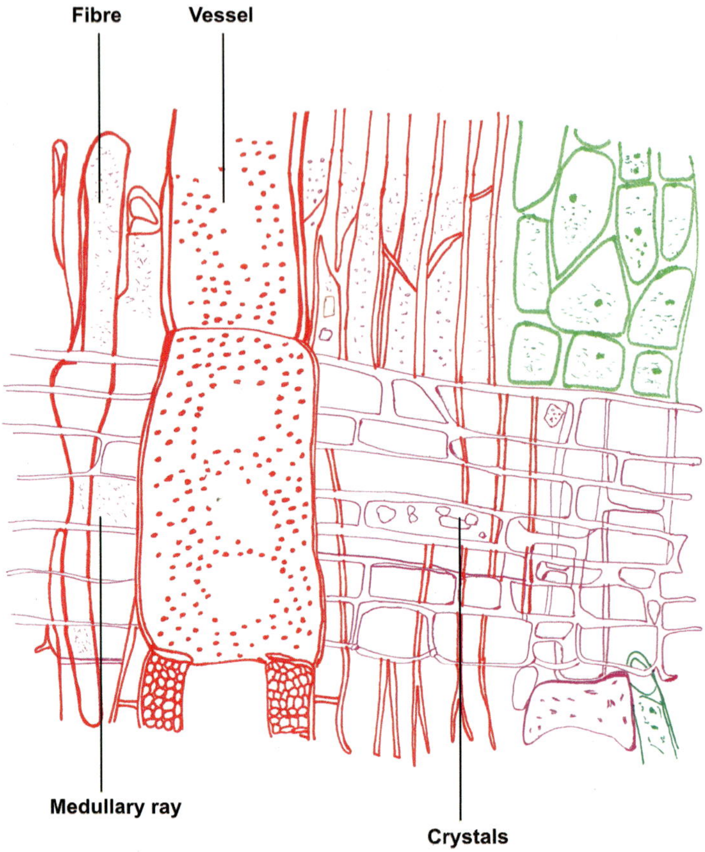

R. L. S. of Quassia wood

Powder Character:

The powder of quassia wood is pale yellowish buff in colour, odourless and intensely bitter in taste.

It shows the following characters microscopically:

a. Vessels:
These are found as fragmented or associated with other xylem tissues. These are numerous which occur singly or in small groups. These vessels are large, bordered and lignified and show numerous minute and bordered pits.
b. Fibres:
These are seen abundantly, which occur in groups and usually found associated with other xylem tissues. These are lignified with moderately thick walls and few pits. These xylem fibres cross the medullary rays at right angles.
c. Medullary rays:
In TLS and RLS medullary rays are mostly multiseriate (three to four celled), and uniseriate rays are sometimes observed. The cells appear polygonal to round in tangential sections with small, numerous pits in tangential walls. The cells are elongated with moderately thickened and lignified wall in radial view. Some cells of medullary rays contain fairly large prisms of calcium oxalate.
d. Calcium oxalate crystals:
A few prisms of calcium oxalate are found scattered as well as within the parenchymatous cells of the xylem and medullary rays. These are of various sizes. Sometimes twin prisms and conglomerate crystals also occur.
e. Starch granules:
These are few, mostly simple and spherical. Occasionally compound grains of two to three components are seen. Hilum is observed as rounded or slit shaped.
f. Xylem parenchyma:
It is generally found associated with xylem vessels and xylem fibres. Xylem parenchymatous cells are thick walled, lignified and heavily pitted. Cells are longitudinally elongated and contain prisms of calcium oxalate.

1. Part of the xylem in R. L. S. showing medullary cells, some containing prisms of calcium oxalate, with underlying fibres and xylem parenchyma
2. Xylem parenchyma
3. Fibres from the phloem
4. Starch granules
5. Part of the xylem in T. L. S. showing uniseriate medullary ray, xylem parenchyma with prisms of calcium oxalate, fibres and part of multiseriate medullary ray
6. Fragments of bordered pitted vessels
7. Xylem fibres
8. Part of the xylem in T. L. S. showing xylem parenchymatous cells, some containing prisms of calcium oxalate, part of medullary ray and bordered pitted vessels
9. Cork in surface view
10. Calcium oxalate prisms

32 Rauwolfia (*Rauwolfia serpentina*)

Synonyms:

Chhota chand, Indian snake root, *Sarpagandha*, Rauwolfia root

Biological Source:

It consists of dried rhizomes and roots of the plant *Rauwolfia serpentina* Benth.

Family:

Apocynaceae

Microscopy:

The transverse section of rauwolfia root shows a circular margin with stratified cork and other features.

The following different tissues are observed microscopically from the periphery towards the centre:

A. Periderm:
It is made up of cork (phellem), phellogen and phelloderm:

i. Cork (phellem):
Cork cells appear isodiametric in surface view. It is typically stratified and consists of two to eight alternating brands of:

a. Smaller, suberised, unlignified radially narrow cells in two to eight layers
b. Larger, suberised, lignified radially broad cells in two to three layers
In most of the commercially available samples of root, cork is exfoliated.

ii. Phellogen:
It is indistinctly seen as a narrow layer of thin-walled cells.

iii. Phelloderm:
It is composed of about 10–12 rows of tangentially elongated to isodiametric cellulosic parenchymatous cells. Cell layers which are away from the phellem show oval cells with intercellular spaces. Some of the cells of phelloderm contain starch grains and calcium oxalate crystals. Starch grains are mostly simple and rarely compound and spherical with triradiate (star shaped) or slit-shaped hilum. Twin prisms of calcium oxalate are also observed.

B. Secondary phloem:
This region is narrow, non-lignified and transversely cut by broad medullary rays. Phloem is composed of sieve tubes, companion cells and phloem parenchyma. Sieve tubes are narrow and scattered. Parenchyma contains starch grains, and some cells of phloem parenchyma contain conglomerate crystals or small prisms of calcium oxalate. Sclerenchymatous cells are absent and if present in the powder indicate admixture with other species like *Rauwolfia tetraphylla*, *Rauwolfia densiflora* and *Rauwolfia micrantha.*

C. Secondary xylem:
Xylem is lignified and usually shows two to six annual rings. The rays of xylem alternate with the medullary rays. Xylem consists of xylem vessels, xylem fibres and xylem parenchyma:

i. Xylem vessels:
Xylem vessels and tracheids are lignified. Vessels are narrow and pitted and a few appear as rounded, polygonal or radially elongated. These are present singly or in the form of group of two to three vessels. Size and number of vessels are less in *Rauwolfia serpentina* as compared to other species. Thus, it can help in detection of adulteration of the genuine sample. Tracheids are pitted with tapering ends.

ii. Xylem fibres:
These appear as rounded or polygonal structures with thick lignified walls. These fibres show pointed or bifurcated ends. These are arranged in tangential bands and radial rows.

iii. Xylem parenchyma:
Cells have lignified and thick walls. Walls show numerous circular pits. Cells contain starch within them. Some cells of xylem parenchyma show calcium oxalate twin prisms.

D. Medullary rays:
These are about one to five cells wide. These rays run radially from the centre through the phloem up to the phelloderm. These are lignified, pitted and large in the xylem area and found prominently uniseriate. These are non-lignified and large in the phloem region. Cells of medullary rays also show starch grains and calcium oxalate crystals.

Rhizome of rauwolfia shows similar characters to those of the root. Along with all other tissues of root, rhizome also contains cortex, pericyclic fibres and small pith. Many of the pericyclic fibres are thick walled and show an ovoid, elongated enlargement near the end. It is a typical feature of the family Apocynaceae. Stone cells are absent in root as well as rhizome.

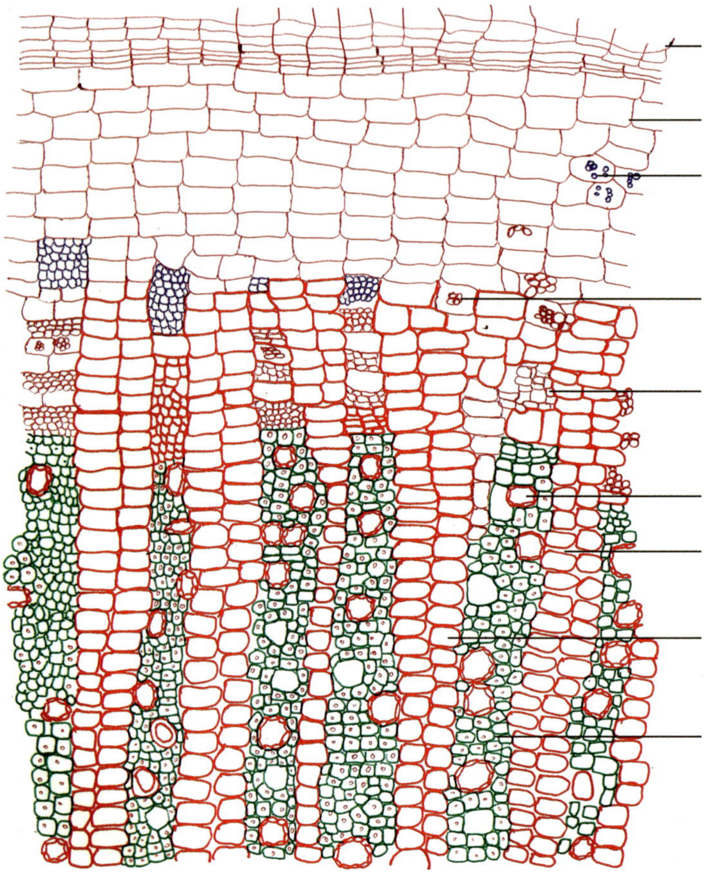

T. S. of Rauwolfia root

Powder Character:

The powder of rauwolfia is pale brownish yellow with faint odour and bitter taste. It shows the following characters:

a. Cork:
   Reddish-brown fragments of cork are seen. Fragments are made up of three to four layers of polygonal, isodiametric, thin-walled stratified cells. Some cork cells are lignified. A few fragments of cork are found attached to the underlying thin-walled phelloderm.
b. Starch granules:
   These are usually contained in the parenchymatous cells. Mostly these are big, simple, spherical and rarely compound (with two to four components) and possess a distinct star or slit-shaped hilum.

c. Parenchyma:
   These cells are lignified and pitted. These parenchymatous cells are from xylem parenchyma and medullary rays. Numerous fragments of parenchyma are observed. The parenchyma cells of medullary rays have moderately thick walls and show rounded to slit-shaped, numerous pits.

   The cells of xylem parenchyma are more elongated and somewhat rectangular in outline, thick walled with numerous pits.
d. Calcium oxalate crystals:
   These are found scattered and in small groups within some of the parenchymatous cells. These are few and in the form of small prisms or conglomerate crystals.
e. Xylem fibres:
   These are rarely seen as isolated or in small groups associated with vessels and tracheids. These are irregular, lignified and thick walled, with small, slit-shaped pits and pointed or bifurcated ends.
f. Vessels and tracheids:
   These occur singly or in small groups. These are narrow, thick, lignified walls having numerous and bordered pits. Vessels are long and few in number with perforated oblique end walls. Tracheids are pitted and have tapering ends.
g. Pericyclic fragments from the rhizome:
   These are found occasionally as large and non-lignified with irregularly thickened walls and show an elongated, ovoid enlargement at one end.
h. Parenchyma of phelloderm and phloem:
   It is seen in small amount. Cells are thin walled and show presence of starch granules. A few cells contain a brownish secretion within them. Cells also show calcium oxalate crystals. Cells of phelloderm have sinuous walls.

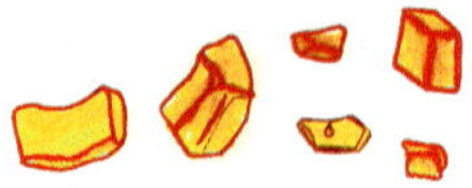

Calcium oxalate crystals

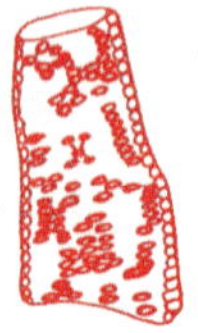

Fragment of bordered pitted vessel

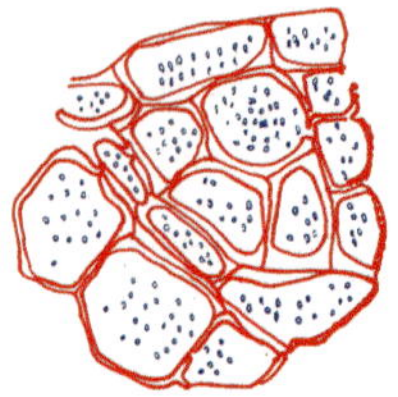

Xylem parenchyma and part of two medullary rays in T. L. S.

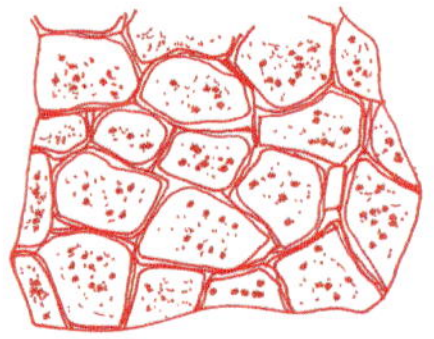

Fragment of medullary ray from the xylem in T. L. S.

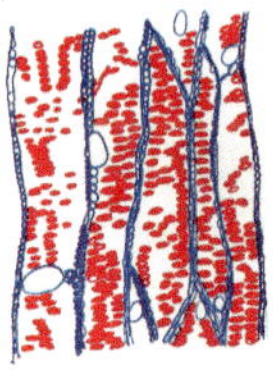

Part of a group of tracheids and tracheidal vessels

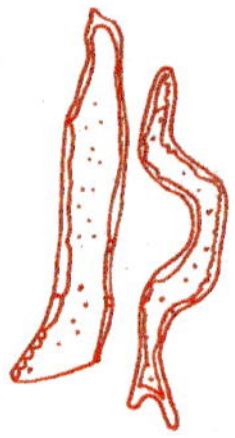

Xylem fibres

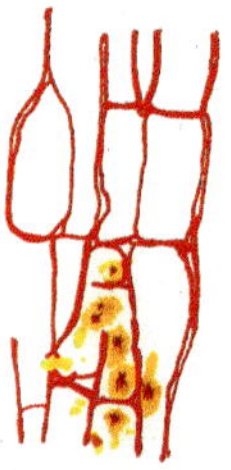

Phloem tissue in T. L. S. with crystals of cacium oxalate

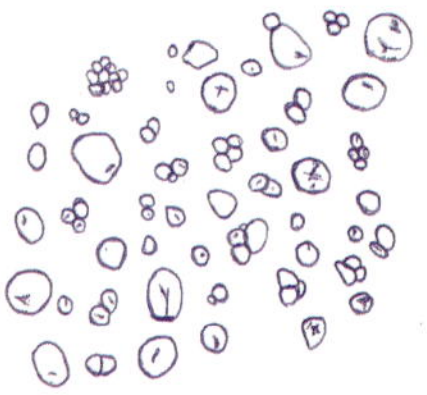

Starch granules

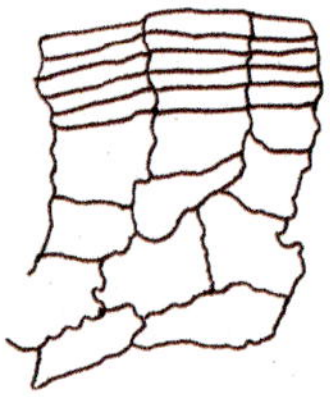

Cork and phelloderm in sectional view

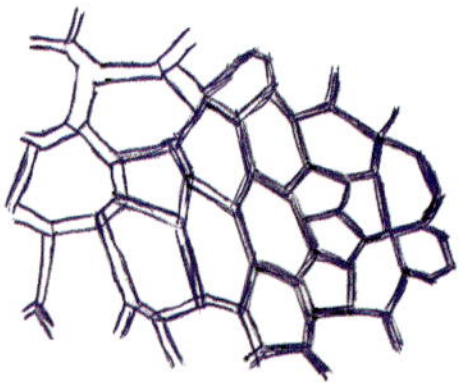

Cork in surface view

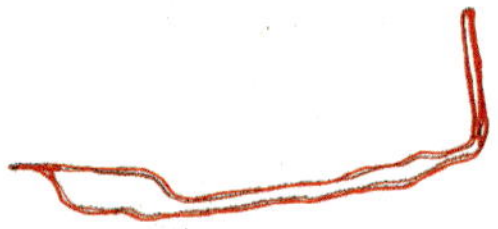

Part of pericyclic fibre from the rhizome

33 Rhubarb (*Rheum officinale*)

Synonyms:

Rhizoma rhei, Radix rhei, Revandchini, Indian rhubarb, Dolu

Biological Source:

It consists of peeled and dried rhizomes and roots of the plant *Rheum emodi* Wall. or *Rheum officinale* Baillon or *Rheum palmatum* Linn.

Family:

Polygonaceae

Microscopy:

The transverse section of rhizome of rhubarb shows the following tissues:

A. Cork:
It is brown in colour and composed of several (10–12) layers of non-lignified rectangular cells.

B. Cortex:
It is broad and occupies the major portion below cork. It is composed of thin-walled, irregular parenchymatous cells. Some of these cells lie within the cells of secondary phloem present below the cortex.

C. Secondary phloem:
It is made up of a few layers of parenchymatous cells.

D. Cambium:
It is wavy and lies between the secondary phloem above and the secondary xylem below. It is compressed many times. The cortex and other tissues within

parenchyma contain starch grains abundantly. Starch grains are simple or compound. Single grains are rounded; components of compound starch grains are muller shaped and hilum is centrally placed. Many parenchymatous cells show cluster crystals of calcium oxalate abundantly. The cortex shows calcium oxalate crystals and starch grains along with tannin masses.

E. Medullary rays:
These are radially placed prominently, consist of one or two layers of cells and contain yellow masses of anthraquinones.

F. Secondary xylem:
It is composed of thick cellulosic walled, reticulate xylem vessels and xylem parenchyma. A continuous ring of star spots is seen just within the inner border of the secondary xylem. The star spots show a small amount of collapsed phloem at the centre, surrounded by phloem developed from the cambium. This cambium arises around the original strand of phloem. Xylem is formed externally around the cambium and composed of large vessels. The medullary rays appear as radiating orange arms of the star. Phloem shows mucilage deposits and cavities originating from the older star spots.

G. Pith:
Pith is composed of thin-walled parenchymatous cells. At the centre of the pith, there are many vertical strands of phloem, and at the nodes there are numerous horizontal strands forming a network. All these strands become encircled by cambia which give rise to concentric bundles having phloem towards the inner central side and xylem on the outer side along with the radiating, slightly curved medullary rays. Therefore, pith shows star spots which are distributed throughout.

There are no sclerenchymatous fibres or cells, and cork is absent in the commercial bark as it has been cut during preparation of the drug for commercial market. Longitudinal section through the xylem region shows xylem vessels with reticulate thickening and elongated cells of xylem parenchyma.

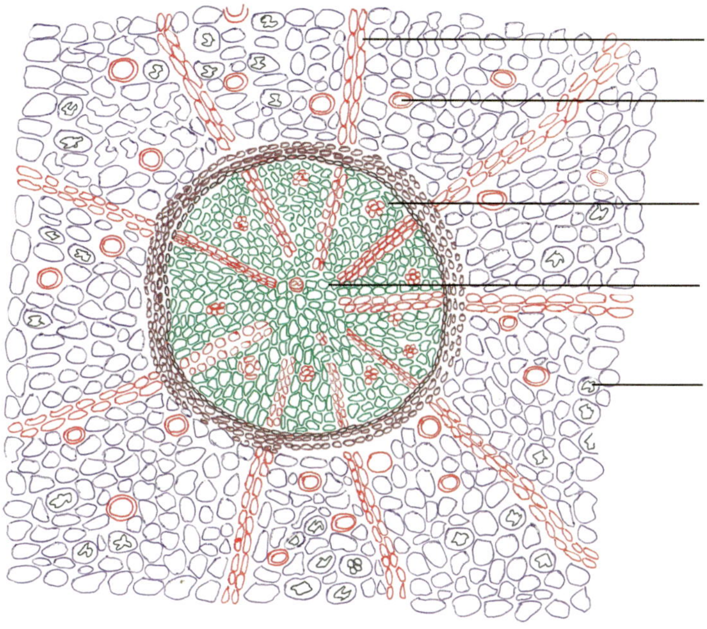

T. S. of Rhubarb rhizome

Powder Character:

The powder of rhizomes of rhubarb is yellowish brown to reddish brown in colour, somewhat gritty with aromatic, characteristic, empyreumatic odour and bitter and astringent taste.

It shows the following characters:

a. Calcium oxalate rosettes:
   These are abundant, scattered in some parenchymatous cells. These are large and sometimes observed as fragmented.
b. Starch granules:
   These are many, simple and spherical or compound with two to five components. These granules have a distinct, central hilum like a cleft or radiating slit.
c. Cork:
   Cork cells are rectangular and thick walled.
d. Vessels:
   Vessels are large and occur singly or in small groups and in the form of fragments. These are reticulately thickened and scattered and do not give reaction for lignin.
e. Parenchyma:
   The parenchyma of medullary rays and ground tissue is abundant. The cells of medullary rays are thick walled and contain yellowish-brown anthraquinone gly-cosidal deposits. Parenchyma of vessels is thin walled, cells elongated at the

ends and filled with starch. The ground tissue is composed of cells with rounded or oval to rectangular margin, containing starch granules or sometimes with large cluster crystals of calcium oxalate; walls are slightly thickened and may show irregular swellings.

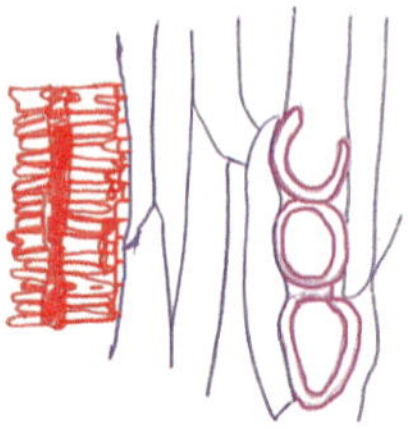

Reticulately thickened vessels associated with parenchyma and a medullary ray in T. L. S.

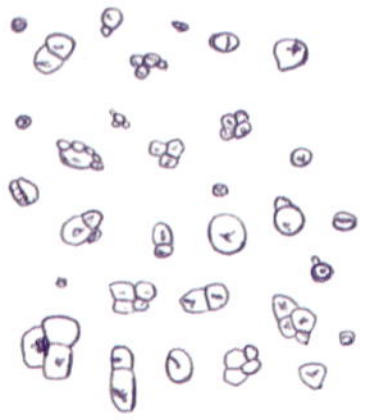

Starch granules

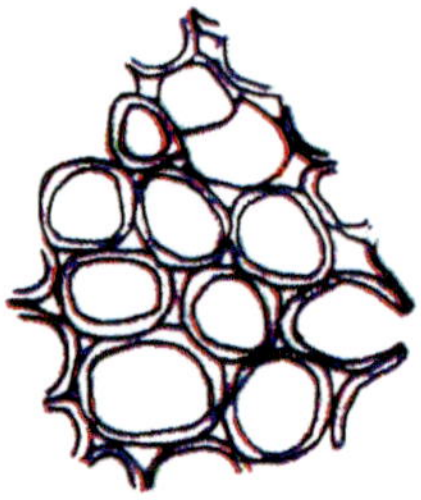

Small-celled, rounded parenchyma of the ground tissue

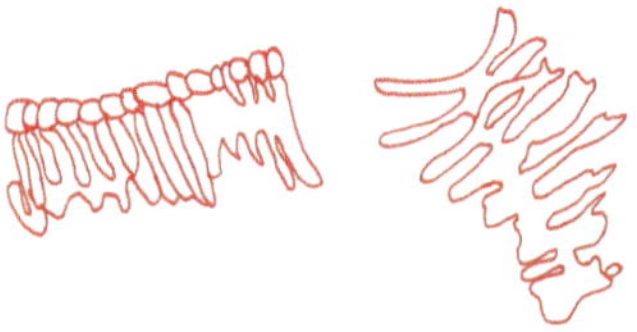

Fragments of reticulately thickened vessels

Cluster crystals of calcium oxalate

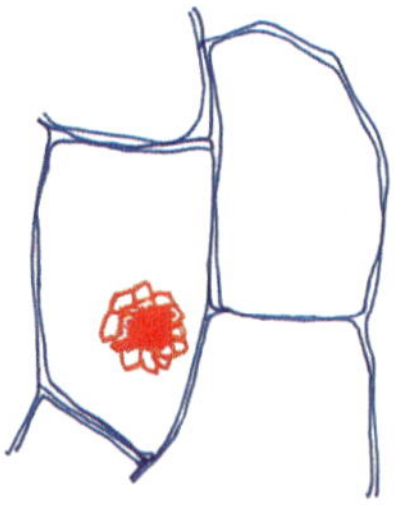

Parenchyma of the ground tissue containing a calcium oxalate crystal

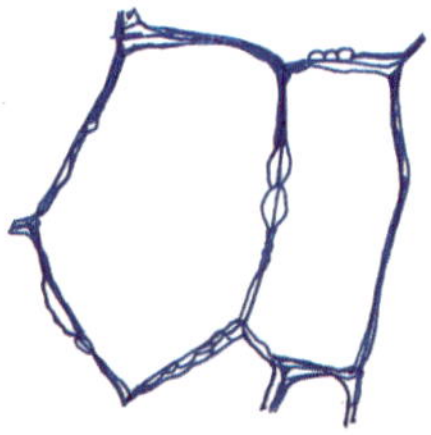

Parenchymatous cells of the ground tissue shwoing irregular thickening

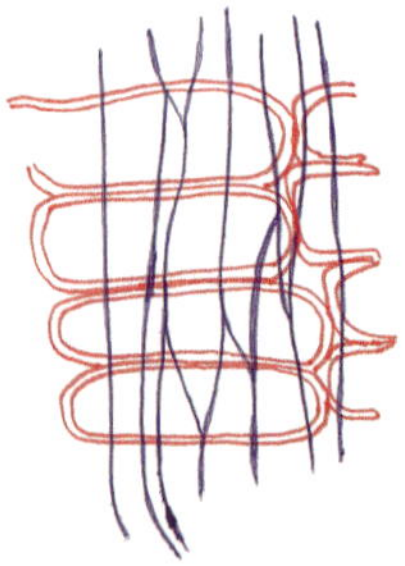

Medullary ray in R. L. S. underlying parenchymatous cells

34 Sandalwood (*Santalum album*)

Synonyms:

Yellow sandalwood, Chandan

Biological Source:

It is scented heartwood obtained from the plant *Santalum album* Linn.

Family:

Santalaceae

Microscopy:

Being wood, all parts of the sandalwood are lignified. Three different sections are taken for anatomical study of sandalwood: transverse section (TS), tangential longitudinal section (TLS) and radial longitudinal section (RLS).

A. Transverse section (TS):
   The transverse section of the wood shows lighter and darker zones of various tissues:

   i. Xylem:
      It mainly consists of vessels and fibres. Vessels are mostly solitary and large and usually occur singly extending from one medullary ray to the next. Sometimes these are seen in small radial groups. Fibres are densely packed with interspersed air spaces (lacunae) and constitute the main bulk of the wood.

ii. Medullary rays:
These are very fine and usually consist of two wide cells closed together. In yellow sandalwood, the volatile oil is deposited in the heartwood and is found in all parts of the wood. It is not secreted by or contained in any gland or particular cells.

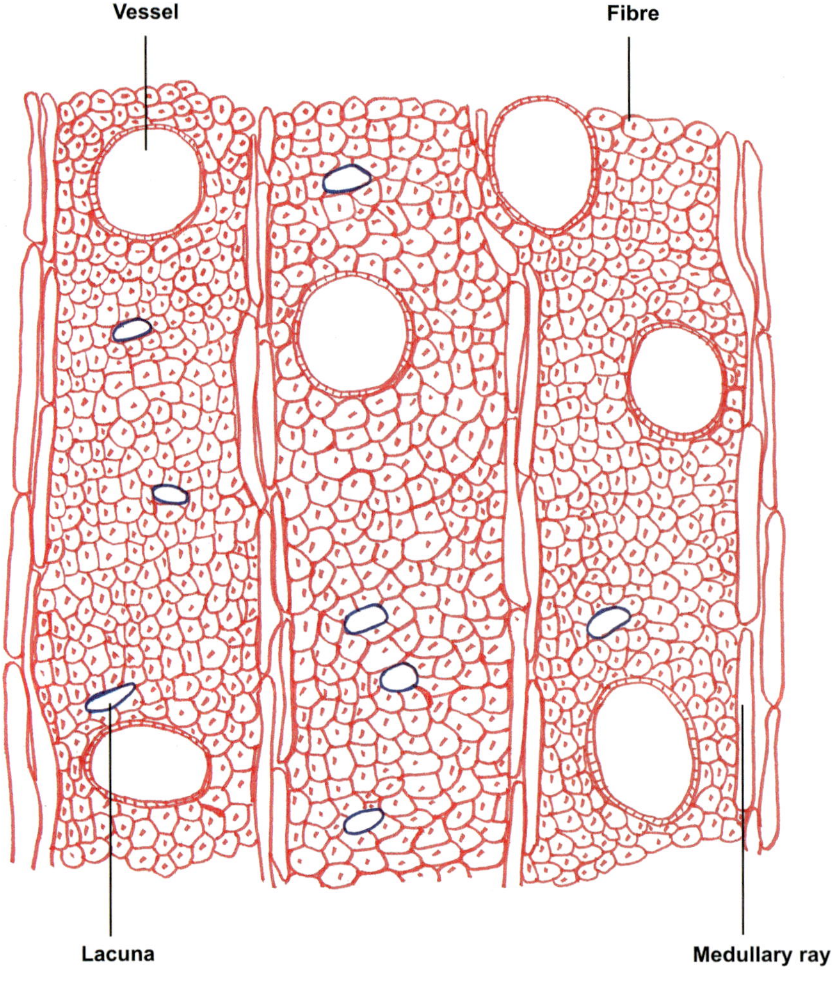

T. S. of Sandalwood

B. Tangential longitudinal section (TLS):
The TLS of the wood shows multiseriate arrangement of medullary rays composed of round ovoid cells. The vessels show minute bordered pits.

C. Radial longitudinal section (RLS):
The RLS shows elongated medullary rays, transversely placed to the vessels and fibres. The walls of ray cell are moderately thickened, and underlying fibres are seen as alternating with vessels.

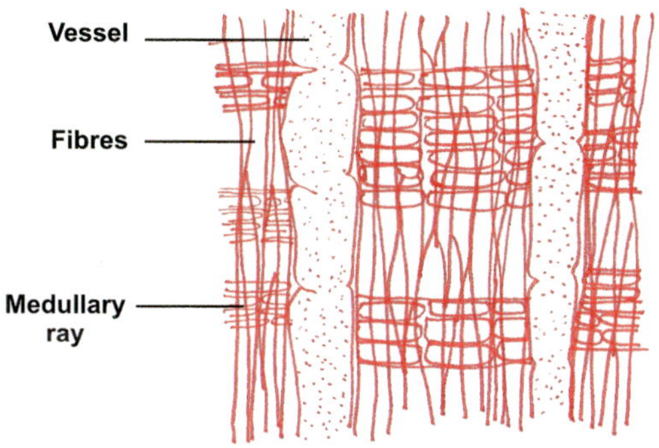

R. L. S. of Sandalwood

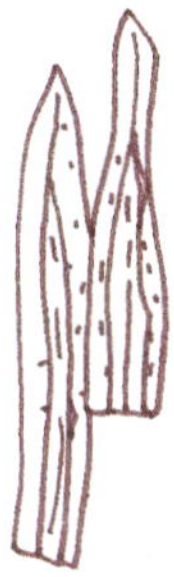

Fibres

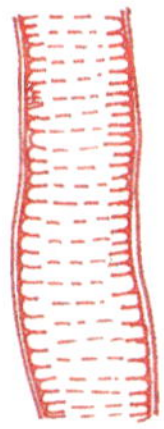

Fragment of vessel

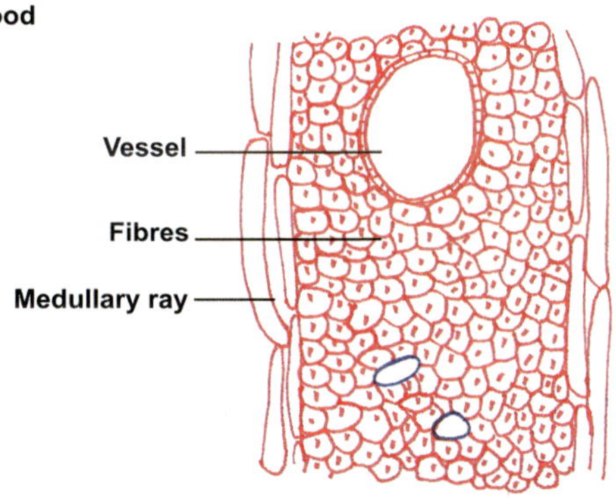

Fragment of wood showing vessel and fibres in sectional view

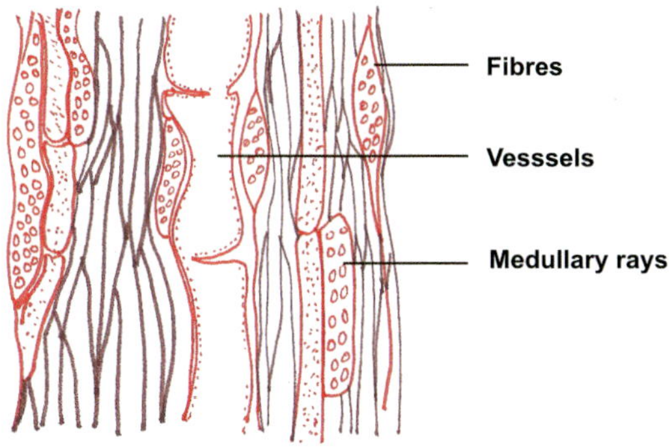

T. L. S. of Sandalwood

Powder Character:

The powder of sandalwood is light brown in colour with strong fragrant characteristic odour and slightly bitter taste. All tissues are lignified. The following characters are seen microscopically:

a. Vessels:
   These are large and wide with a few minute bordered pits. These vessels are found with or without other xylem elements.
b. Fibres:
   These are thick walled, pitted and elongated with fine pointed ends. These fibres occur singly or in groups with interspersed lacunae.
c. Xylem fibres:
   These fibres are sometimes transversed at right angles by thick-walled characteristic medullary ray cells. Oil drops appear to be freely distributed in the wood fibres. A brownish matter is also seen in the wood elements. A few fibres show pitted, thickened walls. Pieces of isolated, thin-walled and non-pitted fibres are also observed as scattered in the powder.
d. Medullary rays:
   These are biseriate and found associated with vessels and fibres.

35 Senna (*Cassia angustifolia*)

Synonyms:

Sonamukhi, Senna ki patti, Indian senna, Tinnevelly senna

Biological Source:

It consists of dried leaflets of the plant *Cassia angustifolia* Vahl. (Indian senna).

Family:

Leguminosae

Microscopy:

The transverse section of a senna leaflet exhibits isobilateral structure under the microscope.

The following tissues are observed in the lamina and midrib region:

A. Lamina:

i. Upper epidermis:
It is composed of polygonal cells arranged in a single layer, covered on the outer side with prominently thick, warty cuticle. Few epidermal cells contain mucilage and straight anticlinal walls. The epidermis bears only non-glandular covering trichomes which are unicellular, short, thick walled, conical, non-lignified, warty and often curved at the bulbous base or with papillose walls. Paracytic stomata are seen at regular intervals.

ii. Mesophyll:
It is differentiated into palisade and spongy parenchyma. Isobilateral structure exhibits presence of upper palisade below the upper epidermis and lower palisade placed above the lower epidermis:

a. Upper palisade:
It is a single layer of elongated, narrow, columnar cells with chloroplastids. The upper epidermis also continues over the midrib region.

b. Spongy parenchyma:
It is made up of loosely arranged parenchymatous cells and contains rosette or prismatic crystals of calcium oxalate.

c. Lower palisade:
It extends to somewhat limited area, i.e. to the lamina region only. Cells are small and loosely arranged and have wavy walls.

d. Lower epidermis:
Cells possessing prominent cuticle and sunken stomata are seen. These cells are somewhat shorter than those of the upper epidermis and have slightly wavy walls. Non-glandular trichomes are also found on the lower epidermis.

B. Midrib:
The transverse section through the midrib region exhibits a flat ventral surface and convex dorsal surface. The epidermal layers are in continuation over the midrib also. The lower epidermis possesses small cells with thick cuticle. The upper palisade is also made up of smaller cells particularly in the midrib region. The lower palisade is absent in the midrib portion, and a group of collenchymatous cells is seen.

At the centre, a group of collateral vascular bundles with xylem on the upper side and phloem beneath is seen. The vascular bundles are covered on both ventral and dorsal sides by an arc of lignified sclerenchymatous fibres. These patches of fibres are somewhat ovate in shape and crescent shaped below. This fibrous arc is characteristic as these fibres are encircled by a layer of parenchyma, with cells of most of it containing prisms of calcium oxalate crystals.

Fibres ensheathed with crystals can be seen occasionally in the lamina portion also. In the surface view, characteristic rubiaceous stomata, covering trichomes and polygonal epidermal cells are seen.

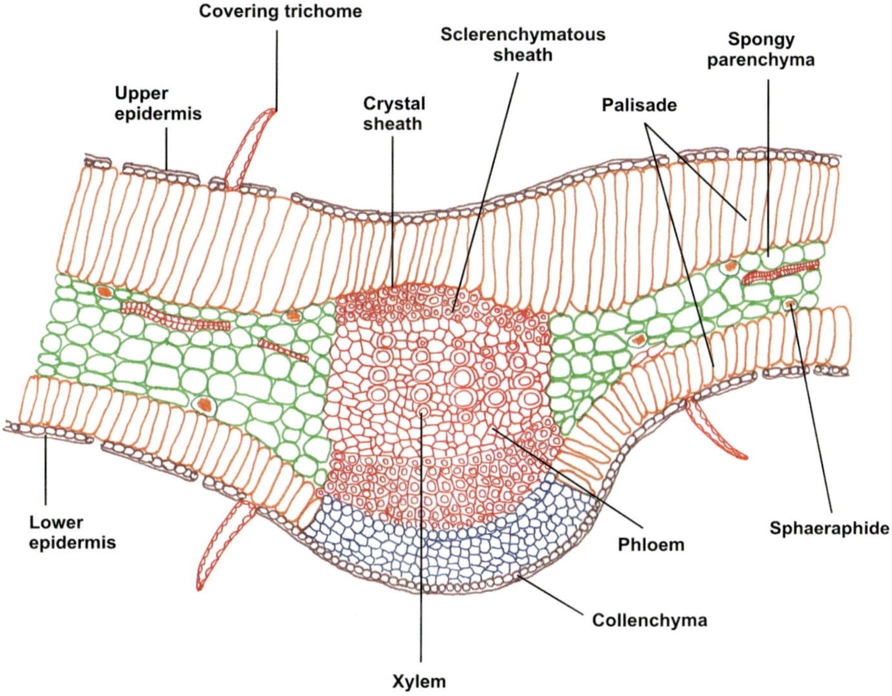

T. S. of Senna leaflet

Powder Character:

The senna leaflet powder is greyish green or yellowish green in colour with a faint, characteristic odour and a mucilaginous, slightly bitter taste.

The powder shows the following features microscopically:

a. Fragments of lamina:
   Upper and lower epidermises of the lamina are similar, cells with thin, straight or slightly sinuous walls and polygonal. Plenty of unicellular trichomes and paracytic stomata are seen. Both epidermises also show cicatrices where trichomes were attached; these consist of small circular scars from which the epidermal cells radiate outwards in a characteristic arrangement.
b. Covering trichomes:
   These are unicellular and conical with thick and warty walls, found attached to pieces of epidermises of the lamina. Straight or curved fragments with thick papillose walls can be seen.

c. Calcium oxalate crystals:
   These are abundant and seen as scattered in the powder. Prisms of calcium oxalate are seen inside the cells of the parenchymatous sheath surrounding the group of sclerenchymatous fibres, and rosettes are seen in the cells of spongy mesophyll. Rosettes (cluster crystals) are of moderate size.
d. Groups of fibres:
   These are thick walled, lignified with few pits and encircled with a sheath of prisms of calcium oxalate.
e. Pieces of lamina in sectional view:
   The palisade cells are seen under the upper epidermis and above the lower epidermis. Palisade cells under the upper epidermis are much elongated and straight walled. Palisade cells above the lower epidermis are short and have sinuous walls. Cells of spongy mesophyll which lie in between these two layers are rounded and contain abundant cluster crystals of calcium oxalate in them. Mucilage is present in many epidermal cells and gives red colour with solution of ruthenium red.
f. Fragments of spiral, annular and pitted vessels are also seen.

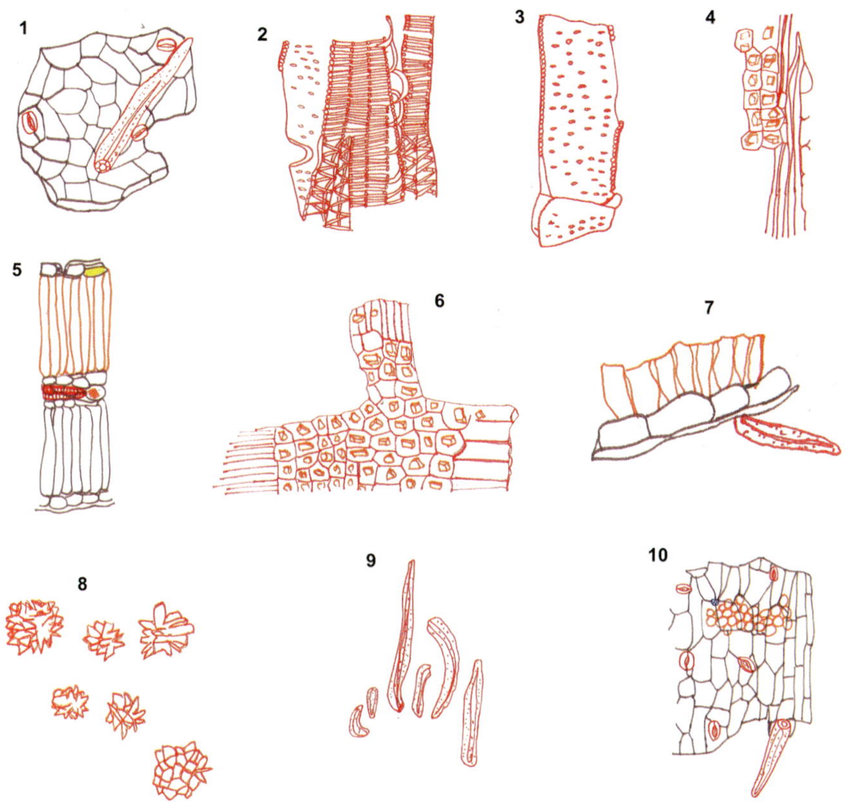

1. Epidermis in surface view with paracytic stomata and an attached trichome
2. Xylem elements from one of the larger veins
3. Part of a pitted vessel from one of the larger veins
4. Part of a group of fibres with calcium oxalate prism sheath
5. Part of the lamina in sectional view showing the upper epidermis containing mucilage, the upper and lower palisade, spongy mesophyll cells containing cluster crystals of calcium oxalate and the lower epidermis
6. Groups of fibres with calcium oxalate prism sheaths at the junction of two small veins
7. Part of the lamina in sectional view with a trichome attached to the lower epidermis
8. Cluster crystals of calcium oxalate
9. Covering trichomes
10. Epidermis in surface view showing paracytic stomata, a cicatrix underlying palisade cells and the elongated cells over a vein with striated cuticle and an attached trichome

36 Squill (*Urginea indica*)

Synonyms:

Indian squill, *Jangli pyaz*

Biological Source:

It consists of dried bulbs and slices of the plant *Urginea indica* Kunth (Indian squill).

Family:

Liliaceae

Microscopy:

Transverse section of scale leaf of squill shows isobilateral pattern with the following tissues:

A. Upper epidermis:
   It is single layered with thick cuticle. Epidermal cells are axially elongated and quadrangular to polygonal in shape, with straight, anticlinal, thin walls. A few anomocytic stomata are seen. Stomata are circular in outline and have wide guard cells. Trichomes are absent.
B. Mesophyll:
   It occupies a major portion and consists of many, thin-walled, large, polyhedral, colourless parenchymatous cells with intercellular spaces. Some of these cells contain bundles of raphides of calcium oxalate embedded in mucilage sheath. A few of mesophyll cells are exceptionally large containing very large calcium oxalate crystals embedded in mucilage. The mucilage sheath is stained red by

corallin soda. Mesophyll cells occasionally show small, rounded starch grains. Small collateral vascular bundles are seen scattered in mesophyll with phloem towards the upper epidermis and xylem towards the lower epidermis.

C. Lower epidermis:
It resembles the upper epidermis. But sometimes lower epidermal cells are larger than those of the upper epidermis, and stomata are less in number.

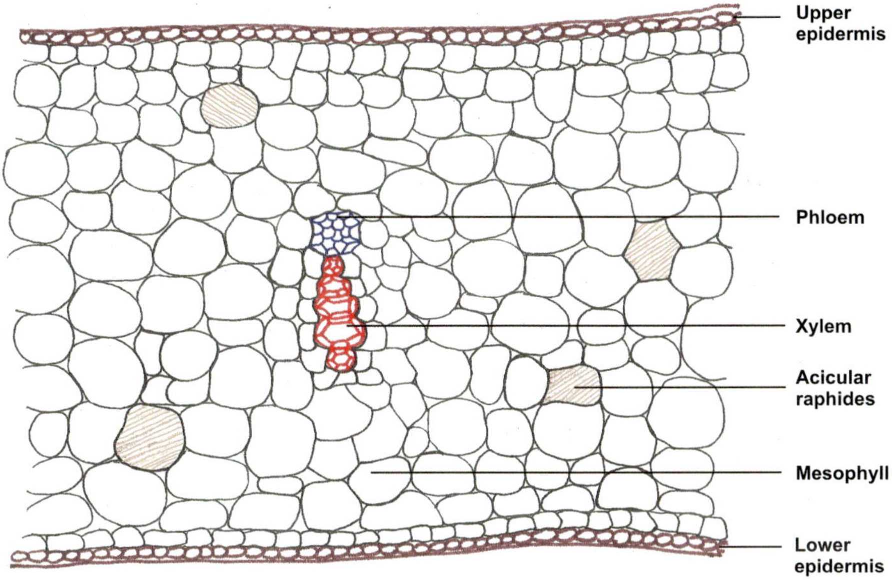

T. S. of slice of Squill bulb

Powder Character:

The powder of squill is off-white to pale buff and very hygroscopic with slight odour and mucilaginous, bitter and acrid taste.

Microscopically it shows the following characters:

a. Acicular crystals of calcium oxalate:
   These are abundant, varying in size, large and single or in bundles, embedded in mucilage within parenchymatous cells. These crystals are also found scattered throughout the powder in broken groups or as single, fragmented forms.
b. Mucilage cells:
   These are seen abundantly either intact along with bundles of acicular crystals of calcium oxalate or in broken fragmented form or with impressions of the crystals. Irregular fragments are found throughout the powder. Mucilage is stained bright red with alkaline solution of corallin and gives a reddish purple colour with solution of iodine.
c. Epidermal cells:
   Fragments of epidermal cells are seen occasionally. These have thin walls, are elongated and have anomocytic stomata. Fragments in sectional view show thick cuticle.
d. Parenchyma:
   It is seen abundantly, composed of thin-walled, rounded to elongated cells with a few, small intercellular spaces. Many cells contain spheroidal masses of crystals of sinistrin (a fructan) and appear pale yellowish in colour. A few groups of slightly thicker-walled parenchymatous cells occur and give a faint reaction for lignin.
e. Vessels:
   These are in small groups or occur singly. Some vessels are large with lignified walls with spiral or annular thickening. Small groups of thin-walled phloem tissue are also found along with these vessels.

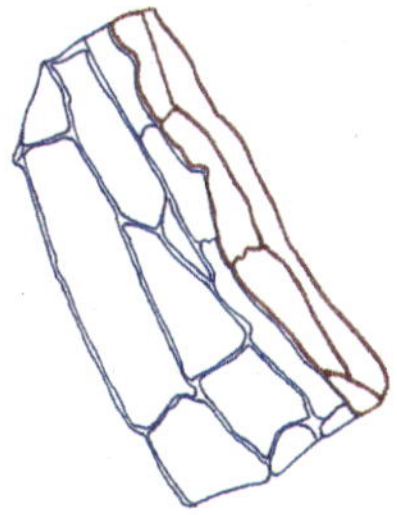

Cuticle, epidermis and parenchyma in sectional view

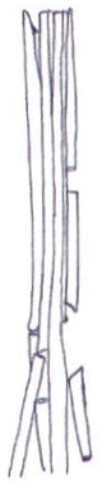

Part of a group of very large acicular crystals of calcium oxalate

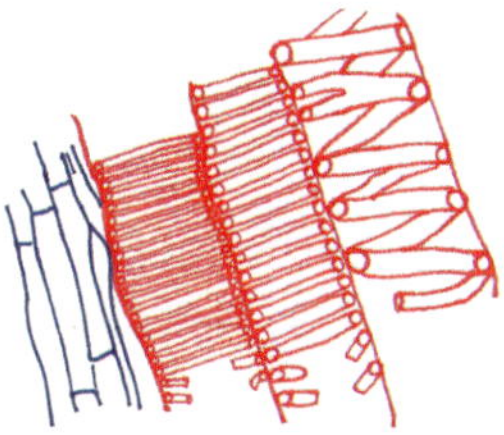

Fragments of vessels with spiral and annular thickening associated with thin-walled phloem tissue

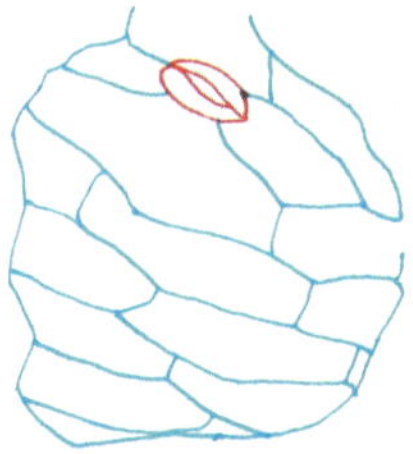

Epidermis in surface view showing a stoma

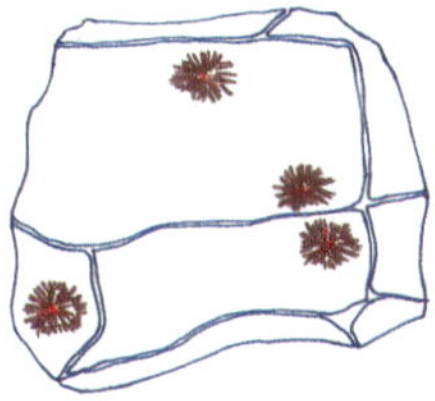

Parenchymatous cells containing spheroidal masses of sinistrin

Fragments of mucilage

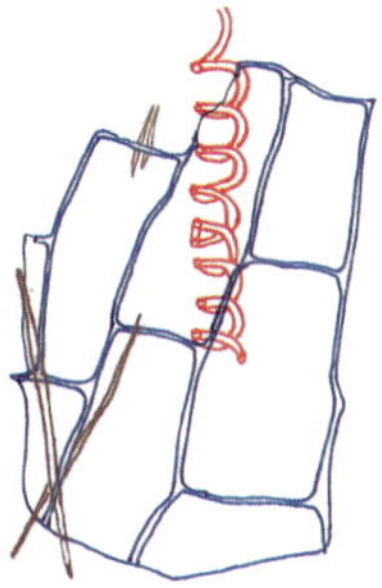

Parenchyma with associated acicular crystals of calcium oxalate and a fragment of a spirally thickened vessel

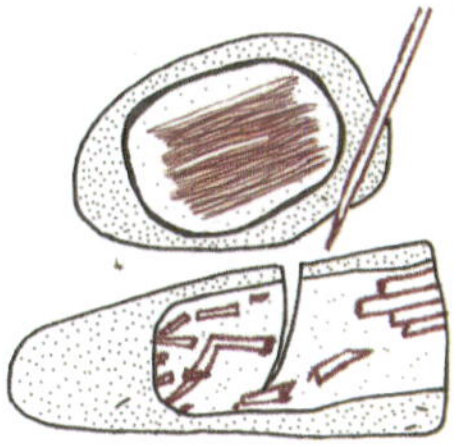

Mucilage cells containing acicular crystals of calcium oxalate

37 Stramonium (*Datura stramonium*)

Synonyms:

Thorn apple, Stramonium leaf, Apple of Peru, *Sada dhatura*, *Safed dhatura*

Biological Source:

It consists of dried leaves and flowering tops of the plant *Datura stramonium* Linn.

Family:

Solanaceae

Microscopy:

Microscopically transverse section of stramonium leaf exhibits the typical dorsiventral pattern.

Transverse section shows the following parts:

A. Upper epidermis:
   Epidermal cells are covered with a smooth cuticle and show wavy walls. This surface shows hairs as well as stomata which are of anisocytic and anomocytic types. Trichomes are of glandular as well as of covering type. Glandular trichomes are small with one- or two-celled stalk and bi- or multicellular (three to seven cells) oval head. Covering trichomes are uniseriate, three to five celled

and slightly curved and have thin warty walls. The length of basal cell is usually more than 50 μm which can be used to distinguish stramonium leaf from *Datura metel* leaf where it is about or less than 35 μm.

B. Palisade:
   A single layer of elongated parenchymatous cells is seen below the upper epidermis.
C. Mesophyll:
   It is made up of a single layer of rectangular cells containing abundant cluster crystals of calcium oxalate. Microsphenoidal and prismatic crystals are found scattered. The remaining part of mesophyll is made up of spongy parenchyma.
D. Midrib:
   It shows bilateral structure and typical subepidermal collenchymas on both surfaces. Sclerenchyma is absent and xylem forms a curved arc.
E. Lower epidermis:
   It is more or less similar to the upper epidermis, but cells particularly have wavy walls.

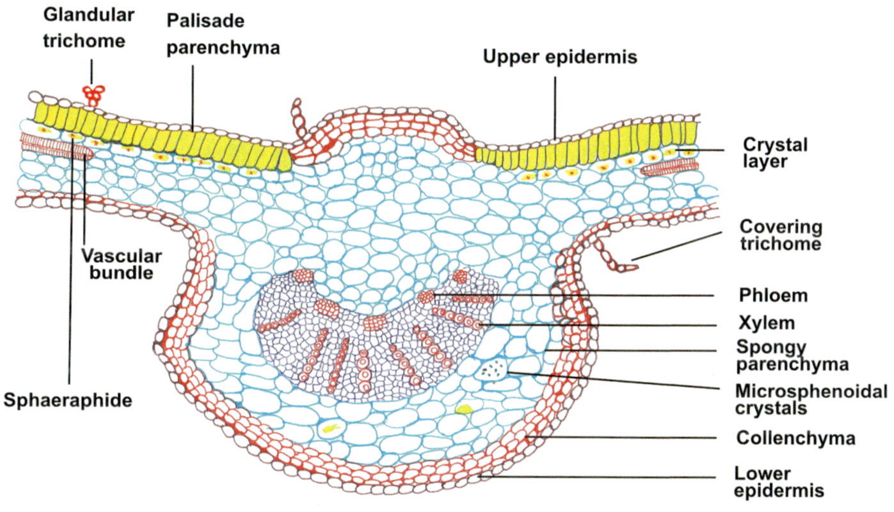

T. S. of Stramonium leaf

Powder Character:

Powdered stramonium leaf is greyish green with faint odour and bitter taste. The powder shows the following diagnostic characters:

a. Fragments of lamina:
   In surface view, abundant fragments of lamina show thin-walled cells (with sinuous outline) of upper epidermis with palisade cells below which are irregular in size and are packed loosely. Cells of the lower epidermis show wavy walls and slight thickening at the corners. The lower epidermis has numerous anisocytic stomata.
b. Trichomes:
   Both covering and glandular trichomes are found abundantly on the upper and lower epidermises. Covering trichomes are uniseriate, three to five celled with conspicuously warty walls. These are conical, wide at base and tapering at the apex. Glandular trichomes have short stalk and oval head composed of three to seven thin-walled cells.
c. Calcium oxalate crystals:
   Cluster crystals of calcium oxalate occur in spongy mesophyll. Crystals are absent from the cells adjacent to the veins. Occasionally prisms of calcium oxalate also occur.
d. Parenchyma:
   It is centrally placed in the midrib region and composed of longitudinally elongated cells with slightly thick walls. Prisms or microsphenoidal crystals of calcium oxalate or cluster crystals of calcium oxalate are seen.
e. Fragments of the lamina in sectional view:
   Fragments of the lamina in sectional view show the tabular epidermal cells with a smooth cuticle. The remaining part of mesophyll is composed of the single layer of palisade cells with the underlying crystal layer and the irregular cells.
f. Pollen grains:
   The occasional fairly large subspherical pollen grains with three pores and an irregularly warted exine are seen.

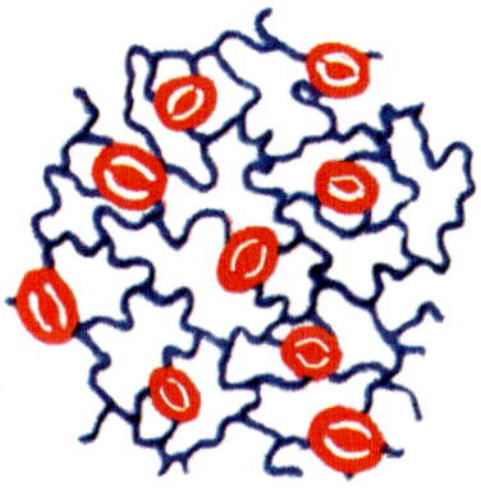

Lower epidermis showing anisocytic stomata

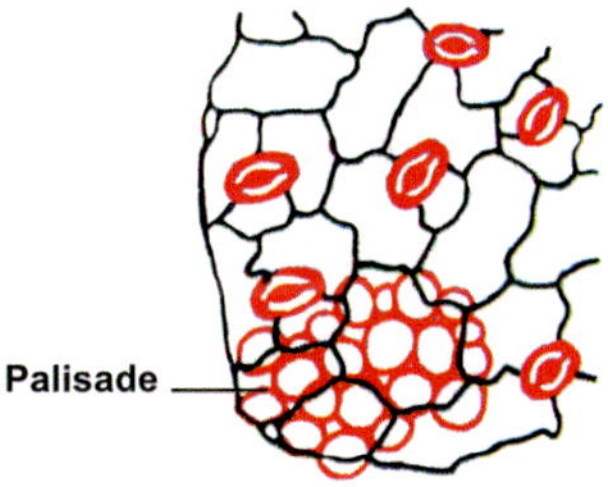

Upper epidermis

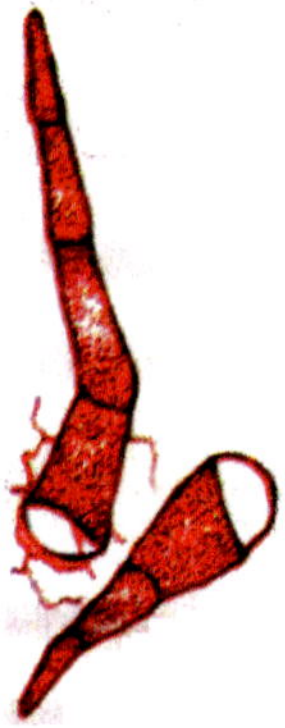

Covering trichomes

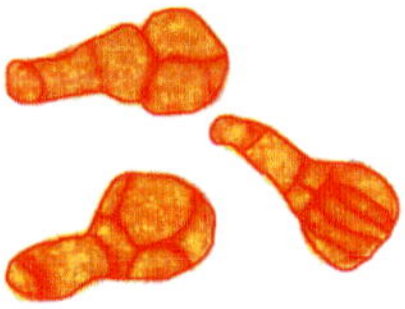

Glandular trichomes

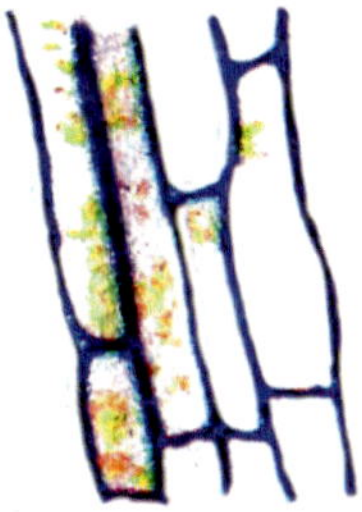

Parenchyma

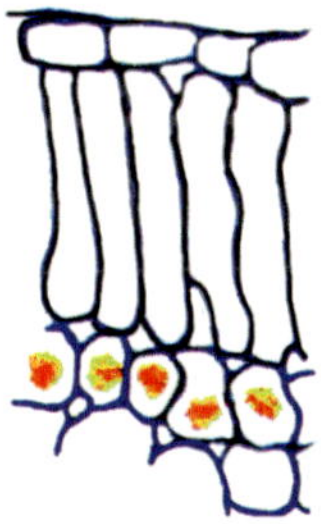

Upper epidermis underlying palisade

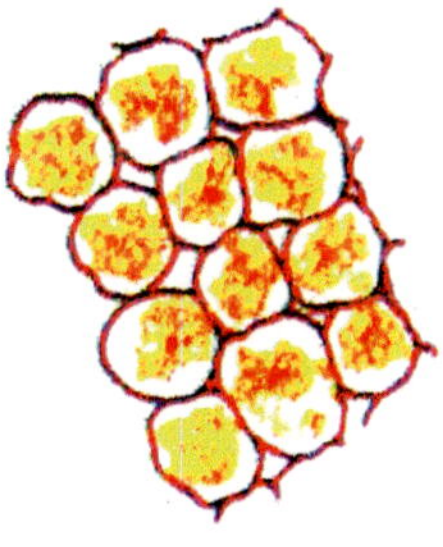

Calcium oxalate crystals

38 Tulsi (*Ocimum sanctum*)

Synonyms:

Sacred basil, Holy basil, Manjari, Krishna tulsi

Biological Source:

It consists of fresh or dried leaves of the plant *Ocimum sanctum* Linn.

Family:

Labiatae

Microscopy:

Transverse section of a tulsi leaf shows a dorsiventral pattern.

The following tissues are observed microscopically in the lamina and midrib region:

A. Lamina:
   It is composed of the upper epidermis, mesophyll and lower epidermis:

   i. Upper epidermis:
      It is composed of a single layer of more or less rectangular cells. The anticlinal walls are sinuous to wavy with thin cuticle. The upper epidermis shows a

few diacytic stomata and numerous trichomes. Trichomes are of both types, i.e. covering and glandular. Covering trichomes occur mainly along the veins. These are long, uniseriate, conical and multicellular (composed of two or three cells and occasionally up to six cells) with slightly thick and warty walls. Glandular trichomes are fairly abundant and typical 'labiate' type with multicellular head. These appear yellowish brown in colour and are of two types. The larger glandular trichomes are sessile with radiate head, composed of eight cells with common cuticle forming a bladder. Some glandular trichomes have a short unicellular stalk and occur in depressions in the epidermis. Each shows a glandular head made up of four radiating cells with a common cuticle slightly raised above to form a spherical bladder-like covering. The glandular trichomes of second type are small and capitate. These show a unicellular stalk and rounded or ovoid head composed of one or two cells. A few glandular trichomes with unicellular stalk and a spherical unicellular head also occur.

ii. Mesophyll:
It is differentiated into palisade and spongy parenchyma:

a. Palisade:
The palisade tissue consists of a single layer of elongated cells below the large and loosely packed upper epidermis.

b. Spongy parenchyma:
It consists of four to six layers of cells with intercellular spaces and oleoresin contents.

iii. Lower epidermis:
It resembles the upper epidermis but possesses diacytic stomata more abundantly. The cell walls are wavier than those of cells of the upper epidermis.

B. Midrib:
The epidermis of the lamina is continuous in the midrib region also. Collenchymatous cells are seen below the upper and above the lower epidermis. The vascular bundle is seen in either of the midrib in which xylem vessels are in the shape of an arc. Phloem tissue is arranged on the dorsal side of the xylem. The trichomes are numerous in the midrib region particularly near the veins.

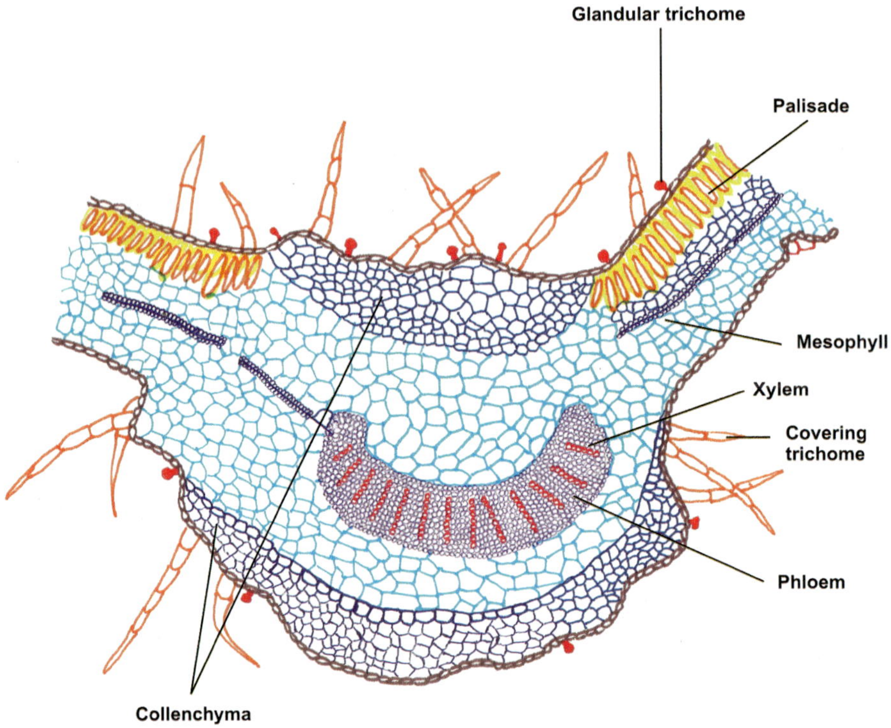

T. S. of midrib of Tulsi leaf

Powder Character:

The powder of tulsi leaves is greyish green with aromatic, strong, characteristic odour and taste.

Microscopically it shows the following characters:

a. Epidermal cells:
   These occur as large and thin walled. The anticlinal walls of cells on the upper epidermis are sinuous to wavy, and those on the lower epidermis are wavier.
b. Palisade cells:
   These appear in sectional view as a single layer of large and loosely packed cells below the upper epidermis. Sometimes spongy parenchyma cells appear along with palisade cells.
c. Trichomes:
   The powder is characterised mainly by presence of numerous glands and trichomes which are of two types, i.e. covering and glandular. Covering trichomes are uniseriate, multicellular and conical, whereas typical 'labiate' type, multicellular heading, glandular trichomes also occur.
d. Stomata:
   These are of diacytic type and more abundant on the lower surface.

e. Vascular tissues:
The powder shows fragments of epidermis with or without trichomes or stomata.

f. Palisade cells, spongy cells, parenchymatous cells with oil glands, collenchymatous cells and xylem vessels with spiral or annular thickening are also seen.

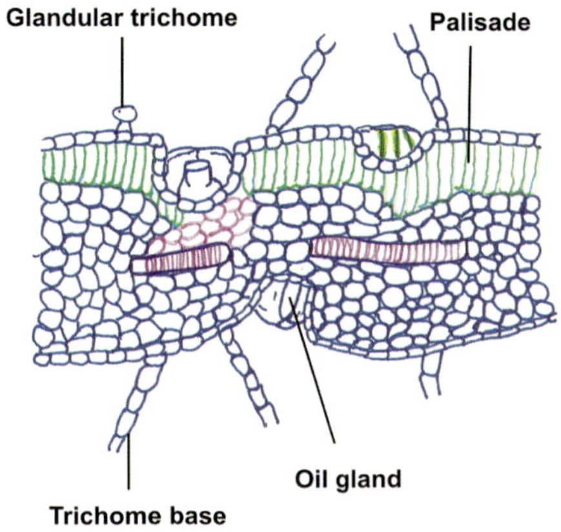

T. S. of lamina

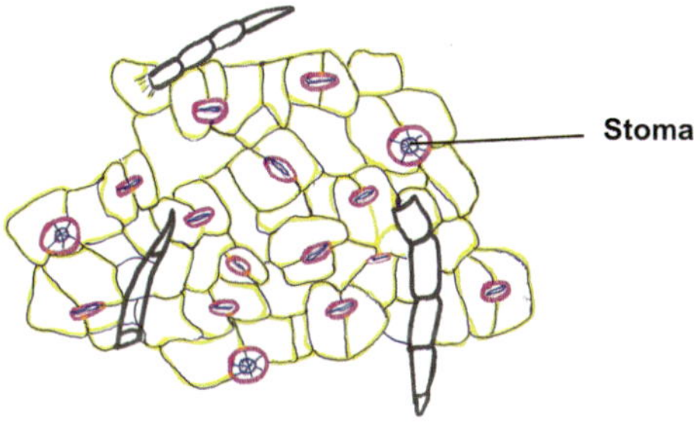

Lower epidermis in surface view

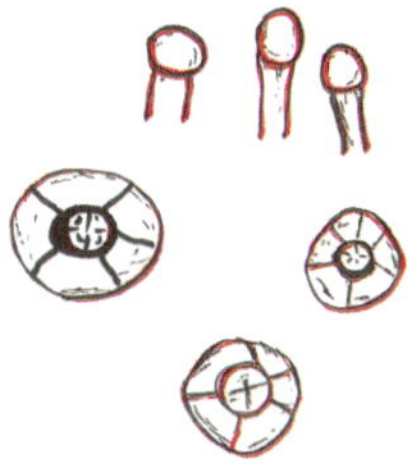

Glandular trichomes

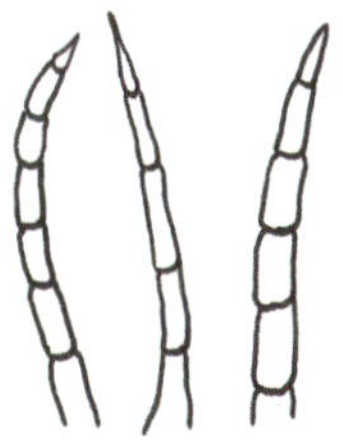

Covering trichomes

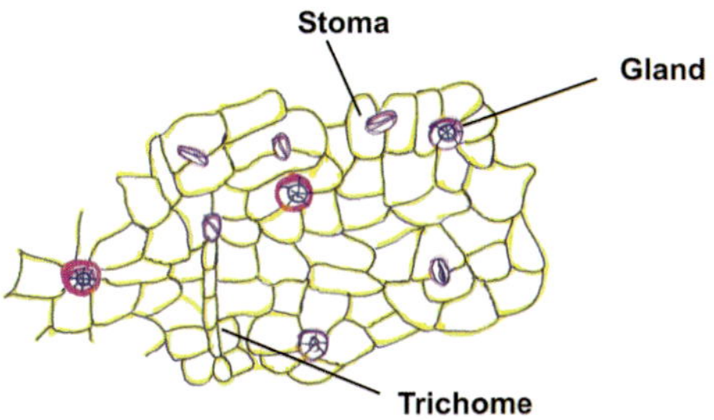

Upper epidermis in surface view

39 Turmeric (*Curcuma longa*)

Synonyms:

Turmeric, Indian saffron, Curcuma, Haldi

Biological Source:

It consists of dried as well as fresh rhizomes of the plant *Curcuma longa* Linn.

Family:

Zingiberaceae

Microscopy:

The powder of dried rhizomes of curcuma is bright yellow with an aromatic, characteristic, pleasant odour. Taste is aromatic, characteristic and pungent.

It shows the following characteristics microscopically:

A. Cork cells:
   Numerous fragments of cork are seen. Cork cells appear pale brown, thin walled, large polygonal and striated in surface view. Fragments exhibit cork as composed of two to five layers of cells and occurring inside the cortex. Sectional view shows cork cells associated with epidermal cells and layers of cortex.
B. Parenchymatous cells:
   Parenchymatous cells are observed abundantly as small groups. These cells are filled with gelatinised starch and bright yellow colouring matter. On clearing, cells are seen as round to oval in shape and possess thin, irregular walls.

C. Starch granules:
   These are simple, flattened, oblong and oval, with a small pointed hilum and few faint, transverse striations. The starch is mainly gelatinised and appears as big rounded, pasty masses with yellowish tinge.
D. Covering trichomes:
   These are few, distinct, unicellular, elongated and conical. These are seen as bluntly pointed, with thick striated walls, and enlarged bases show pitted walls. These trichomes are found scattered and sometimes found as attached to the pieces of epidermis.
E. Vessels:
   These are abundantly seen as large, well developed and wide. These are mainly reticulately thickened with regular, rectangular pits, but a few vessels also show spiral and annular thickening.
F. Oleoresin cells:
   These are seen as scattered all over.
G. Epidermis:
   It is composed of a layer of cells which are tabular and polygonal to elongated in surface view. Thin walls are straight, slightly thick and pitted. The epidermis also shows presence of rounded stomata, cicatrices and covering trichomes.
H. Fragments are not easily detected in the powder being very much indistinct and small.

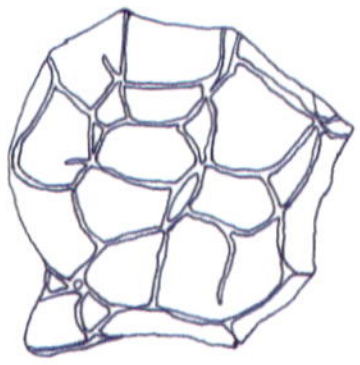

Parenchymatous cells

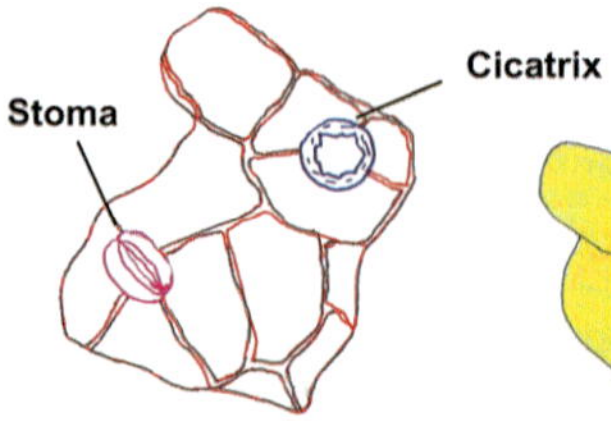

Epidermis in surface view showing a stoma and a cicatrix

Parenchymatous cells filled with gelatinised starch and yellow colouring matter

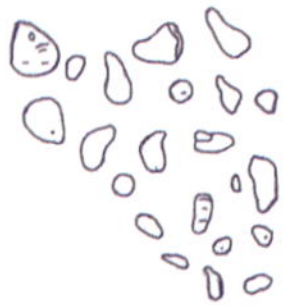

Starch granules

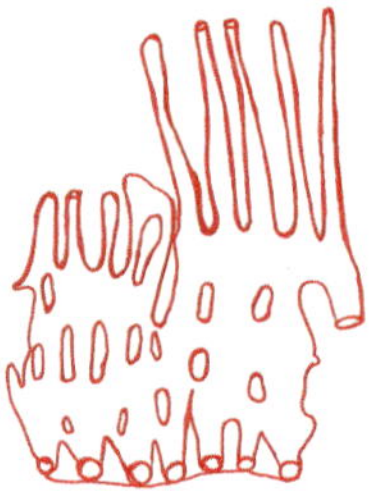

Fragment of a reticulately thickened vessel

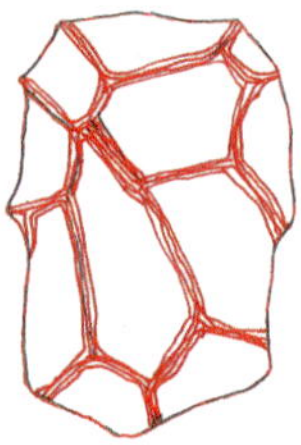

Cork in surface view

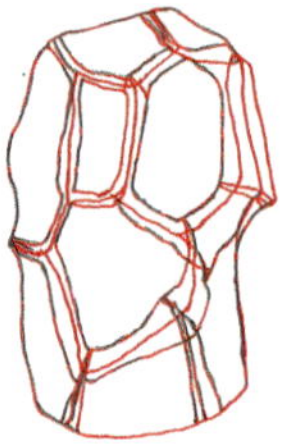

Cork in oblique surface view

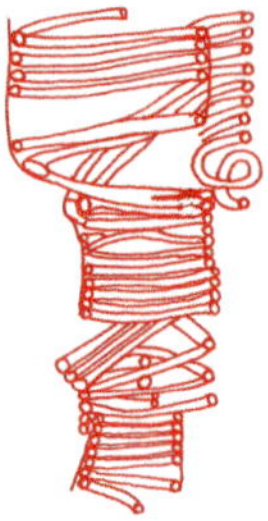

Fragments of spirally thickened vessels

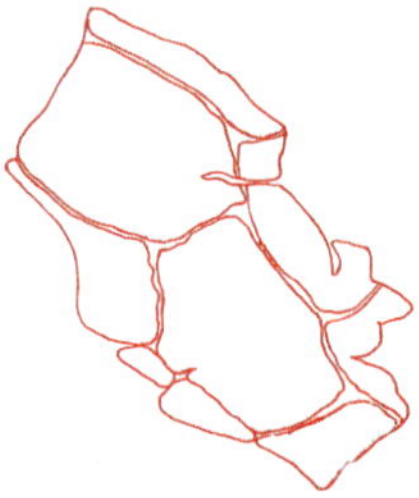

Epidermis in surface view

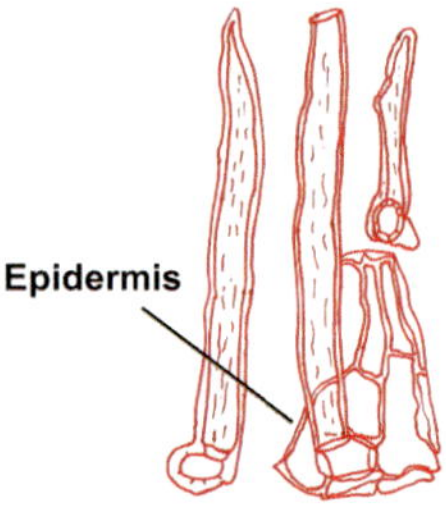

Covering trichomes, one attached to a fragment of the epidermis

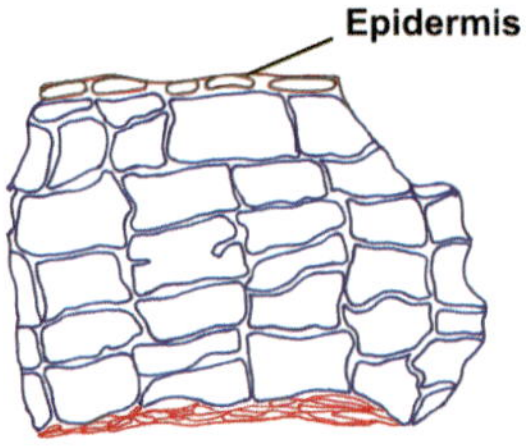

Part of the outer tissues in sectional view showing the epidermis, cortex, cork and a layer of collapsed cells

40 Vasaka (*Adhatoda vasica*)

Synonyms:

Adhatoda, Adulsa, Malabar nut, Acusha

Biological Source:

It consists of fresh and dried leaves of the plant *Adhatoda vasica* Nees.

Family:

Acanthaceae

Microscopy:

The transverse section of a vasaka leaf exhibits dorsiventral pattern microscopically. The following tissues are observed in the lamina and midrib region:

A. Lamina:
   It is composed of the upper epidermis, mesophyll and lower epidermis:

   i. Upper epidermis:
      It is seen as a single layer of thinly cuticularised cells. It shows a wavy outline in surface view. Cells are more or less rectangular with wavy walls. Both covering and glandular trichomes are observed as emerging from the upper epidermal cells. These trichomes are more abundant near the veins. Covering trichomes are uniseriate, two to four celled, warty, conical, bent,

thick walled and pointed. Glandular trichomes are small and sessile and have a quadricellular head. Diacytic (caryophyllaceous) type of stomata occurs mostly near the veins.

ii. Mesophyll:
It can be differentiated into (a) upper layer as palisade and (b) lower layer as spongy parenchyma:

a. Palisade:
Two layers of palisade cells are observed in the transverse section. Cells are radially elongated and compactly placed. A few cylindrical cystoliths can be seen in this region after mounting the section in water.

b. Spongy parenchyma:
It is made up of three to six layers of loosely arranged irregular cells with intercellular spaces. Many veins pass through the spongy region. These layers also show presence of vascular bundles structurally similar as in the midrib region.

iii. Lower epidermis:
It resembles the upper epidermis structurally, but cells are less wavy than those of the upper epidermis. Trichomes and stomata are more in number than the upper epidermis.

B. Midrib:

i. Epidermis:
The epidermal cells of the lamina are continuous over the midrib also. These cells are seen as cubical and heavily cuticularised with numerous trichomes.

ii. Collenchyma:
It is placed below the upper epidermis and above the lower epidermis. It is composed of about two to six layers of collenchymatous cells.

iii. Cortical parenchyma:
The remaining region of the midrib is occupied by cortical parenchyma with about three to five vascular bundles. Amongst these the central one is the largest. Cortical parenchyma shows cystoliths, oil globules and calcium oxalate crystals.

iv. Vascular bundles:
These are seen as collateral and arc shaped. The lignified xylem is on the ventral side, and non-lignified phloem is on the dorsal side. These tissues are transversed by a few radiating medullary rays.

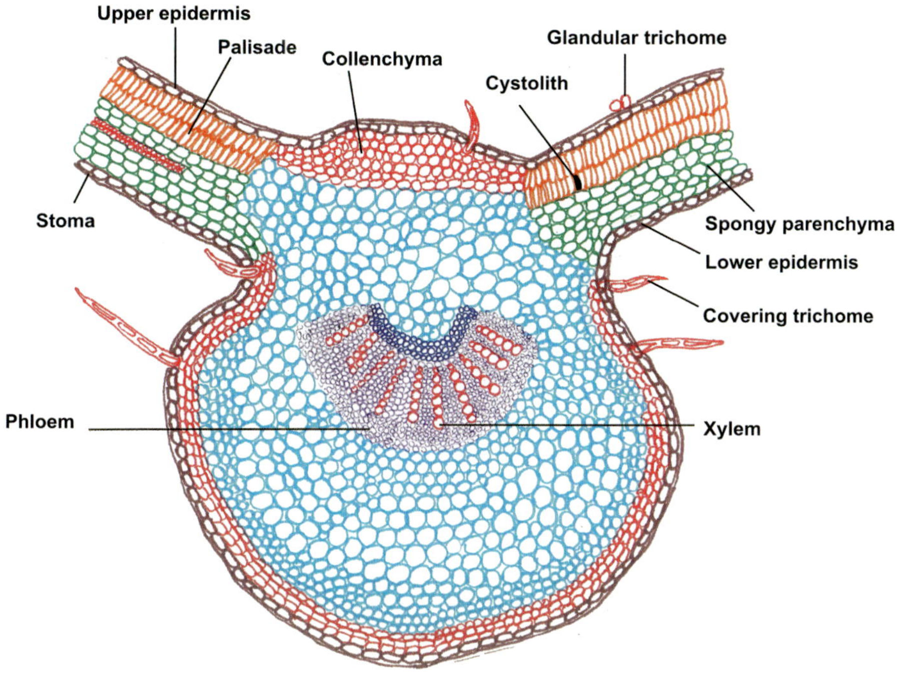

T. S. of Vasaka leaf

Powder Character:

The powder of the vasaka leaves is greyish brown with faint odour and intensely bitter taste. It shows the following characters microscopically:

a. Epidermal cells:
   Numerous fragments of the epidermis as well as whole cells are seen. Cells have wavy anticlinal walls, trichomes and diacytic stomata.
b. Trichomes:
   Both covering and glandular trichomes are seen. Covering trichomes are warty, conical and uniseriate and have one to three cells. Glandular trichomes are small and sessile with quadricellular head.
c. Stomata:
   Numerous diacytic stomata are seen on the upper as well as lower epidermis.
d. Palisade cells:
   Fragments or double-layered palisade cells are seen.
e. Xylem vessels:
   Vessels are lignified along with spiral and annular thickening.
f. Cystoliths:
   Mesophyll shows presence of a few cylindrical cystoliths. These cell contents get dissolved in hydrochloric acid with effervescence.
g. Calcium oxalate crystals:
   Acicular and prismatic forms of calcium oxalate crystals are observed.

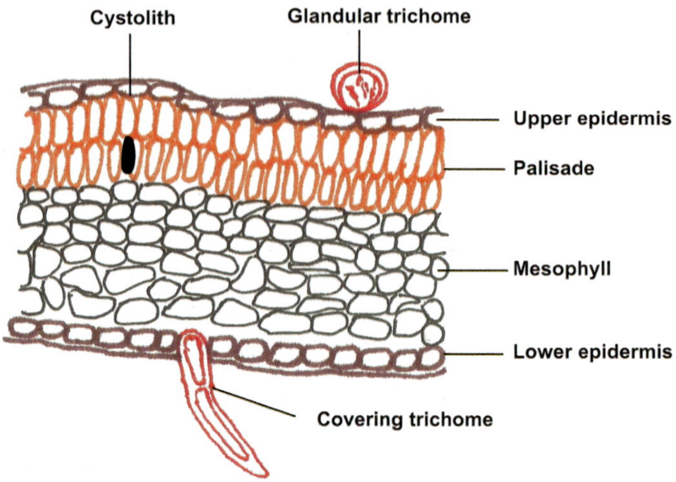

T. S. of Lamina

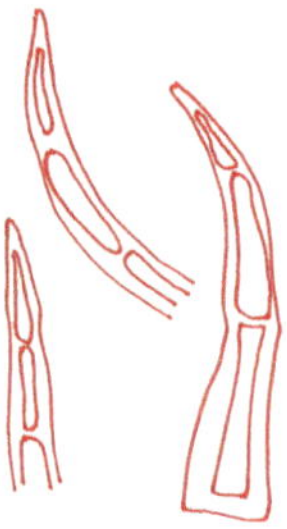

Covering trichomes

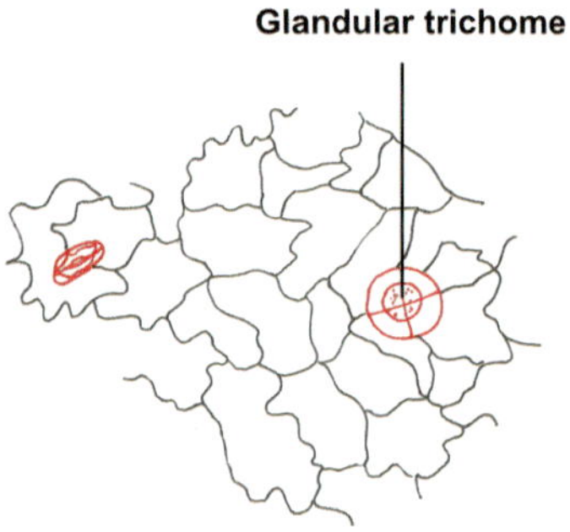

Upper epidermis in surface view

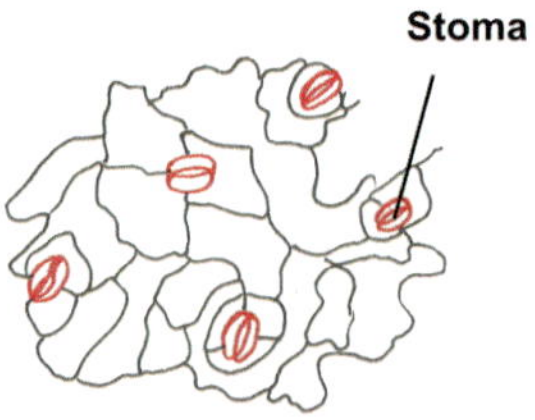

Lower epidermis in surface view

Glandular trichomes

41 Vinca (*Catharanthus roseus*)

Synonyms:

Catharanthus, Madagascar periwinkle, Sadabahar, Barmasi, Lochnera rosea

Biological Source:

It consists of dried leaves of *Catharanthus roseus* G. Don.

Family:

Apocynaceae

Microscopy:

Microscopically a vinca leaf shows a typical dorsiventral structure.

The following tissues are observed in the transverse section in the lamina and midrib region:

A. Lamina:

   It is composed of the upper epidermis, mesophyll and lower epidermis:

   i. Upper epidermis:

      It is made up of a single layer of more or less rectangular cells, the outer wall of which is cuticularised. Anticlinal walls are straight on the upper surface whereas wavy on the lower surface. Anisocytic stomata appear on the upper epidermis. Covering trichomes emerge which are unicellular, long and warty and has a bulbous base. A few very short trichomes are also seen occasionally.

   ii. Mesophyll:

It can be differentiated into palisade and spongy parenchyma:

a. Palisade:
   A single layer of elongated and compactly placed palisade cells is seen below the upper epidermis.
b. Spongy parenchyma:
   It is composed of five to eight layers of loosely arranged cells with intercellular spaces. Vascular strands are observed. Calcium oxalate crystals of any type are totally absent.

iii. Lower epidermis:
   It resembles the upper epidermis, but cell walls are wavy and stomata are more in number on the lower surface.

B. Midrib:
   Epidermal layers of lamina are continuous in the midrib region also. Collenchymatous strips appear below the upper epidermis as well as above the lower epidermis. These are composed of thick-walled cellulosic cells. The remaining region is occupied by cortical parenchyma along with well-developed zone of collateral vascular bundles at the centre of the midrib. Xylem is lignified with spiral and pitted vessels and phloem is non-lignified.

   The surface preparation of the leaf shows cruciferous (anisocytic) stomata and covering trichomes.

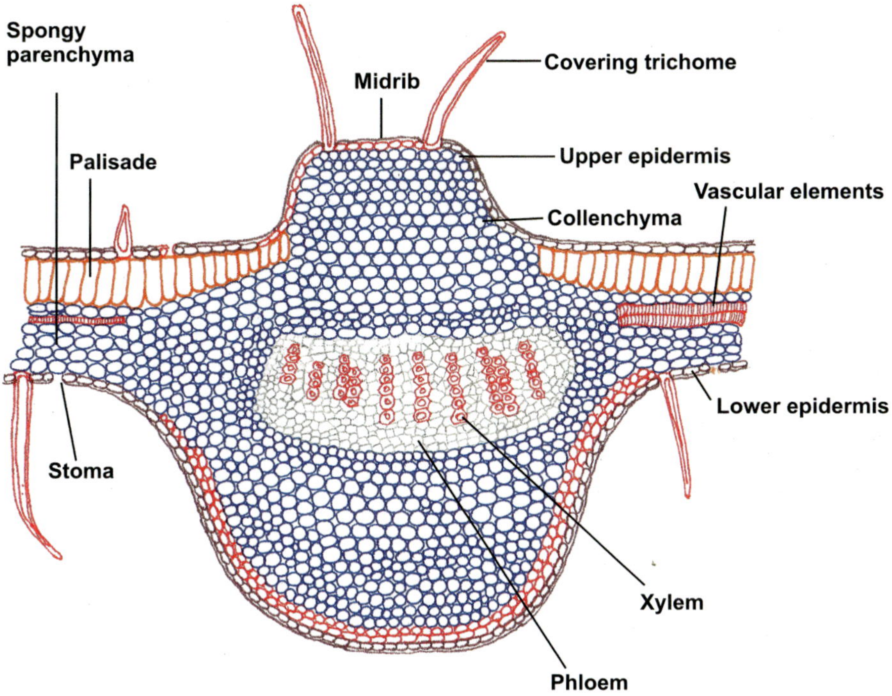

T. S. of Vinca Leaf

Powder Character:

The powder of vinca leaves is dark green in colour with faint odour and slightly bitter taste.

It shows the following characters:

a. Epidermal cells:
   These are seen as whole or as pieces of rectangular, thin, straight or wavy anticlinal walled cells. Occasionally epidermal cells with stomata or trichomes appear in powder.
b. Trichomes:
   Covering trichomes are observed which are unicellular, dagger shaped and warty with bulbous base.
c. Mesophyll:
   Palisade and spongy parenchyma with cells of upper epidermis are seen.
d. Stomata:
   Anisocytic (three subsidiary cells, one is smaller than other two) stomata are observed on both epidermises.
e. Vessels:
   Spiral and pitted vessels and collenchymatous cells can be seen along with cortical parenchyma.

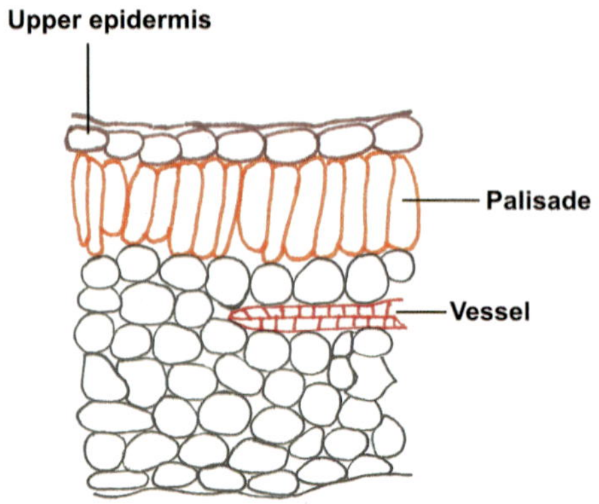

T. S. of part of lamina

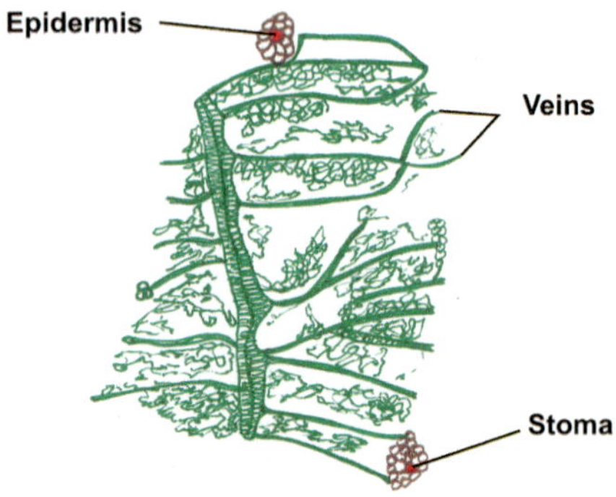

Portion of leaf in surface view showing veins and epidermis

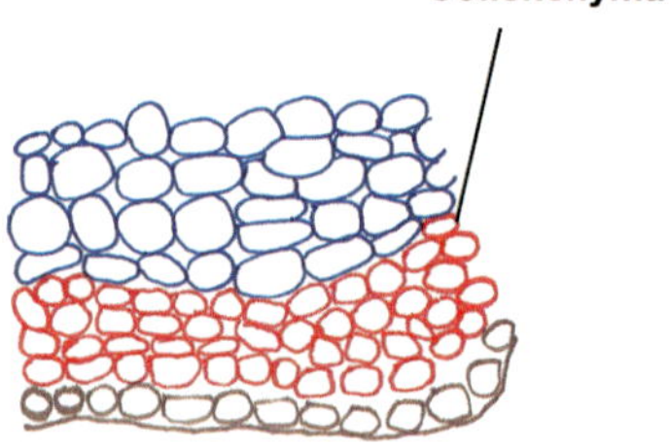

T. S. of part of leaf

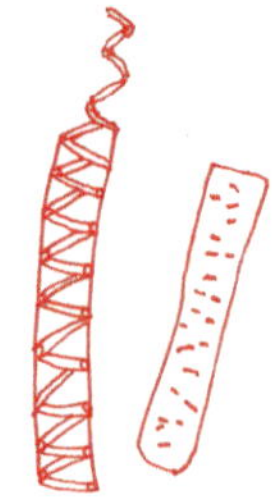

Vessels

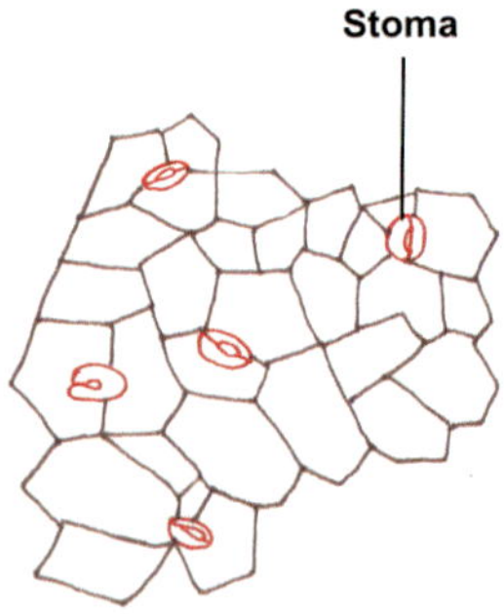

Upper epidermis in surface view

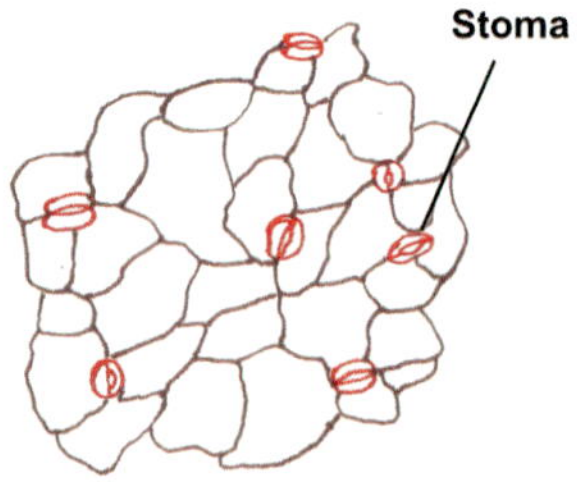

Lower epidermis in surface view

Bibliography

Bendre A (2007) Practical botany. Rastogi Publications, Meerut
Datta C, Mukerji B (1952) Pharmacognosy of Indian leaf drugs, Pharmacognosy Laboratory Bulletin No. 2. Government of India, Calcutta
Iyengar M (1997) Pharmacognosy of powdered crude drugs, 5th edn. Manipal Power Press, Manipal
Iyengar M, Nayak S (1998) Anatomy of crude drugs, 7th edn. Manipal Power Press, Manipal
Jackson B, Snowdon D (1992) Atlas of microscopy. CBS Publications, Delhi
Khandelwal K (2008) Practical pharmacognosy. Nirali Publications, Pune
Kokate C (1996) Practical pharmacognosy. Vallabh Publications, Delhi
Serrano R, Silva G, Silva O (2010) Application of light and scanning electron microscopy in the identification of herbal medicines. In: Méndez-Vilas A, Díaz J (eds) Microscopy: science, technology, applications and education. Formatex, Madrid, pp 182–190
Trease G, Evans W (2006) Pharmacognosy, 15th edn. Bailliere Tindall, London
Vasudevan T, Laddha K (eds) (2003) Herbal drug microscopy. Yucca Publications, Mumbai
Wallis T (2005) Textbook of pharmacognosy. CBS Publications, Delhi
Yeung E (1998) A beginner's guide to the study of plant structures. In: Karcher SJ (ed) Proceedings of the 19th workshop/conference of the Association for Biology Laboratory Education (ABLE). Vol 19 Tested studies for laboratory teaching. ABLE, pp 125–142
Zhongzhen Z (2010) Application of microscopic techniques for the authentication of herbal medicines. In: Méndez-Vilas A, Díaz J (eds) Microscopy: science, technology, applications and education. Formatex, Madrid, pp 803–812

S.S. Gurav and N.S. Gurav (eds.), *Indian Herbal Drug Microscopy*,
DOI 10.1007/978-1-4614-9515-4,

Index

A

Acanthaceae, 189
Aconite. *See Aconitum napellus*
Aconitum napellus
  biological source, 15
  family, 16
  microscopy
    endodermis, 16
    metaderm, 16
    parenchyma, 17
    pith, 17
    primary cortex, 16
    primary phloem, 17
    secondary phloem, 17
    xylem, 17
  powder characters
    fibres, 19
    outer layer, fragments of, 19
    parenchyma, 17
    sclereid, 19c
    starch granules, 19
    vessels, 19
Adhatoda vasica
  biological source, 189
  family, 189
  microscopy
    lamina, 189–190
    midrib, 190
  powder characters
    calcium oxalate crystals, 191
    cystoliths, 191
    epidermal cells, 191
    palisade cells, 191
    stomata, 191
    trichomes, 191
    xylem vessels, 191
Aloe. *See Aloe barbadensis*
Aloe barbadensis
  biological source, 21
  family, 21
  microscopy
    epidermis, 21
    mucilaginous parenchyma, 21–22
    palisade, 21
    parenchyma, 21
    vascular bundles, 22
Anise. *See Pimpinella anisum*
Apocynaceae, 113, 155, 193
Arjuna. *See Terminalia arjuna*
Ashoka. *See Saraca indica*
Ashwagandha. *See Withania somnifera*
Asparagus. *See Asparagus racemosus*
Asparagus racemosus
  biological source, 43
  family, 43
  microscopy
    cortex, 44
    endodermis, 44
    epidermis, 43
    exodermis, 44
    pericycle, 44
    pith, 44
    vascular bundles, 44
  powder characters
    calcium oxalate crystals, 46
    epidermal cells, 46
    parenchyma, 46
    xylem vessels, 46
Atropa belladonna
  biological source, 48
  family, 48
  microscopy

S.S. Gurav and N.S. Gurav (eds.), *Indian Herbal Drug Microscopy*,
DOI 10.1007/978-1-4614-9515-4,

Atropa belladonna (*cont.*)
cambium, 49
periderm, 49
phelloderm, 49
secondary phloem, 49
secondary xylem, 49
powder characters
calcium oxalate, 51
cork cells, 51
parenchyma, 51
starch granules, 51
xylem parenchyma, 51
xylem vessels and fibres, 51
Ayurvedic Formulary of India, 2
Azadirachta indica
biological source, 128
family, 128
microscopy
lamina, 128–129
midrib region, 129
powder characters
cortical cells, 130
epidermis, 130
palisade cells, 130
pith cells, 130
spongy parenchyma, 130
trichomes, 130
vascular tissues, 130
vessels, 130

B
Belladonna. *See Atropa belladonna*
Berberidaceae, 143

C
Cannabaceae, 53
Cannabis. *See Cannabis sativa*
Cannabis sativa
biological source, 53
family, 53
microscopy
lamina, 54
midrib, 54
powder characters
bracteoles, surface view, 55
calcium oxalate crystals, 55
fragments of bracts, 55
stigmas, 55
trichomes, 55
Caraway. *See Carum carvi*
Carum carvi
family, 58
microscopy
cremocarp, 58
embryo, 59
endocarp, 59
endosperm, 59
epicarp, 59
testa, 59
powder characters
endocarp, 59
endosperm, 59
epicarp, 59
fibrovascular tissue, 61
sclereids, 61
testa, 61
vittae, 60
Cassia angustifolia
biological source, 169
family, 169
microscopy
lamina, 170
midrib region, 170–171
powder characters
calcium oxalate
crystals, 172
groups of fibres, 172
lamina, 170
trichomes, 171
Catharanthus roseus
biological source, 193
family, 193
microscopy
lamina, 193–194
midrib region, 194
powder characters
epidermal cells, 195
stomata, 195
trichomes, 195
vessels, 195
Cephaelis ipecacuanha
biological source, 104
family, 104
microscopy
cambium, 105
cortex, 105
periderm, 104
phellogen, 104
phloem, 105
xylem, 105
powder characters
calcium oxalate crystals, 106
cork, 106
fibrous cells, 107
parenchyma, 107
sclereids, 107

starch granules, 106
tracheids, 107
Cinchona. *See Cinchona calisaya*
Cinchona calisaya
biological source, 63
family, 63
microscopy
cortex, 64
medullary rays, 65
periderm, 64
phloem fibres, 64–65
phloem parenchyma, 64
sieve tubes, 64
powder characters
calcium oxalate crystals, 66
cork cells, 66
fibres, 66
parenchyma, 66
starch grains, 66
stone cells, 66
Cinnamomum zeylanicum
biological source, 68
family, 68
microscopy
cortex, 68
medullary rays, 69
pericycle, 69
pericyclic fibres, 69
periderm, 68
phloem fibres, 69
phloem parenchyma, 69
sclereids, 69
powder characters
calcium oxalate crystals, 71
cork cells, 71
fibres, 71
oil cells, 71
sclereids, 71
starch grains, 71
Cinnamon. *See Cinnamomum zeylanicum*
Clove. *See Eugenia caryophyllus*
Colchicum. *See Colchicum autumnale*
Colchicum autumnale
biological source, 78
family, 78
microscopy
endosperm, 79
pigment layer, 79
starch granules, 78
strophiole, 79
testa, 79
Combretaceae, 28
Crude drugs, 2
Curcuma longa
biological source, 186
family, 186
microscopy
cork cells, 186
epidermis, 187
oleoresin cells, 187
parenchymatous cells, 186
starch granules, 187
trichomes, 187
vessels, 187
Cytochemical stains, 13

D

Datura stramonium
biological source, 179
family, 179
microscopy
mesophyll, 179
midrib, 179
palisade, 179
upper epidermis, 178
powder characters
calcium oxalate crystals, 180
fragments of lamina, 180
lamina, 180
parenchyma, 180
pollen grains, 180
trichomes, 180
Digitalis. *See Digitalis purpurea*
Digitalis purpurea
biological source, 81
family, 81
microscopy
lamina, 81–82
midrib, 82
powder characters
fragments of lamina, 84
parenchyma, 84
trichomes, 84

E

Ephedra. *See Ephedra gerardiana*
Ephedraceae, 86
Ephedra gerardiana
biological source, 86
family, 86
microscopy
cortex, 87
epidermis, 87
pericyclic fibres, 87

Ephedra gerardiana (*cont.*)
pith, 87
vascular bundles, 87
powder characters
brownish matter, 88
epidermal cells, 88
fibres, 88
wood elements, 88
Eucalyptus. *See Eucalyptus globulus*
Eucalyptus globulus
biological source, 90
family, 90
microscopy
collenchyma, 91
lamina, 91
vascular bundles, 91
powder characters
calcium oxalate crystals, 92
cell contents, 92
epidermis, 92
pericyclic fibres, 92
stomata, 92
xylem vessels, 92
Eugenia caryophyllus
biological source, 73
family, 73
microscopy
columella, 74
cortex, 74
epidermis, 73–74
powder characters
aerenchyma, 76
calcium oxalate crystals, 76
fibres, 76
hypanthium, 76
oil glands, 76
parenchyma, 76
pollen grains, 76
sclereids, 76
starch grains, 76

F

Fennel. *See Foeniculum vulgare*
Foeniculum vulgare
biological source, 94
family, 94
microscopy
cremocarp, 94–95
embryo, 95
endocarp, 95
endosperm, 95
epicarp, 95
mesocarp, 95
testa, 95
powder characters
endocarp, 96
endosperm, 97
epicarp, 96
fibrovascular tissue, 97
mesocarp, 96
vittae, 97
Freehand sectioning methods, 6–7

G

Ginger. *See Zingiber officinale*
Glycyrrhiza glabra
biological source, 123
family, 123
microscopy
periderm, 123–124
pith, 124–125
secondary phloem, 124
secondary xylem, 124
powder characters
calcium oxalate crystals, 126
cork cells, 126
fibres, 126
parenchyma, 126
starch grains, 126
vessels, 126
Good Manufacturing Practices (GMP), 1

H

Herbal drug microscopy
aconite, 15–20
aloe, 21–22
anise, 23–27
arjuna, 28–32
ashoka, 33–37
ashwagandha, 38–42
asparagus, 43–47
belladonna, 48–52
cannabis, 53–57
caraway, 58–62
cinchona, 63–67
cinnamon, 68–72
clove, 73–77
colchicum, 78–80
digitalis, 81–85
ephedra, 86–89
eucalyptus, 90–93
fennel, 94–98
ginger, 99–103

ipecacuanha, 104–108
ispaghula, 109–112
kurchi, 113–116
linseed, 117–122
liquorice, 123–127
neem, 128–131
nutmeg and mace, 132–134
nux vomica, 135–139
opium, 140–142
podophyllum, 143–147
quassia, 148–154
rauwolfia, 155–160
rhubarb, 161–164
sandalwood, 165–168
senna, 169–173
squill, 174–177
stramonium, 178–181
tulsi, 182–185
turmeric, 186–188
vasaka, 189–192
vinca, 193–196
Herbal drugs
authentication, 1–2
microscopic techniques, 2
use of, 1
Histological and histochemical staining techniques
cytochemical stains, 13
iodine–potassium iodide, 11–12
microscopic observation, powdered drugs, 13–14
phloroglucinol–HCl, 11
Sudan dyes, 12–13
TBO, 10
Holarrhena antidysenterica
biological source, 113
family, 113
microscopy
cortex, 114
periderm, 113
secondary phloem, 114
powder characters
calcium oxalate crystals, 116
cork cells, 116
medullary rays, 116
phloem fibres, 116
starch grains, 116
stone cells, 116

I
Indian Ayurvedic Pharmacopoeia, 2
Iodine–potassium iodide, 11
procedure, 12
results, 12
stain preparation, 12
Ipecacuanha. *See Cephaelis ipecacuanha*
Ispaghula. *See Plantago ovata*

K
Kurchi. *See Holarrhena antidysenterica*

L
Labiatae, 182
Lauraceae, 68
Leguminosae, 33, 123, 169
Liliaceae, 21, 43, 78, 174
Linaceae, 117
Linseed. *See Linum usitatissimum*
Linum usitatissimum
biological source, 117
family, 117
microscopy
cotyledons, 111
endosperm, 110
testa, 118–119
powder characters
endosperm and cotyledons, 111
hyaline layer, 121
parenchyma, 121
pigment layer, 120
sclerenchyma, 120
testa, 120
Liquorice. *See Glycyrrhiza glabra*
Loganiaceae, 135

M
Meliaceae, 119, midrib region119
Microscopy
aconite, 16–17
aloe, 21–22
anise, 23–25
arjuna, 28–29
ashoka, 33–34
ashwagandha, 38–39
asparagus, 43–44
belladonna, 48–49
cannabis, 54
caraway, 58–59
cinchona, 64–65
cinnamon, 68–69
clove, 73–74
colchicum, 78–79
digitalis, 81–82
ephedra, 87

Microscopy (*cont.*)
eucalyptus, 90–91
fennel, 94–95
ginger, 99–100
ipecacuanha, 104–105
ispaghula, 109–110
kurchi, 113–114
linseed, 118–119
liquorice, 123–125
neem, 128–129
nutmeg and mace, 132–133
nux vomica, 135–136
opium, 140–141
podophyllum, 143–144
rauwolfia, 155–157
rhubarb, 161–162
sandalwood, 165–167
senna, 170–171
squill, 174–175
stramonium, 178–179
tulsi, 182–183
turmeric, 186–187
vasaka, 189–190
vinca, 193–194
Myristicaceae, 132
Myristica fragrans
biological source, 132
family, 132
microscopy
endosperm, 133
perisperm, 132–133
powder characters
cell contents, 133
endosperm, 133
parenchyma, 133
perisperm, 133
starch grains, 133
Myrtaceae, 73, 90

N
Neem. *See Azadirachta indica*
Nutmeg and mace. *See Myristica fragrans*
Nux vomica. *See Strychnos nux-vomica*

O
Ocimum sanctum
biological source, 182
family, 182
microscopy
lamina, 182–183
midrib region, 183
powder characters
epidermal cells, 184
palisade cells, 184
stomata, 184
trichomes, 184
vascular tissues, 184
Opium. *See Papaver somniferum*

P
Papaveraceae, 140
Papaver somniferum
biological source, 140
family, 140
microscopy, 140–141
Phloroglucinol–HCl
procedures, 11
results, 11
stain preparation, 11
Picrasma excelsa
biological source, 148
family, 148
microscopy
radial longitudinal section, 148
tangential longitudinal section, 148
transverse section, 148–149
powder characters
calcium oxalate crystals, 154
fibres, 154
medullary rays, 154
starch granules, 154
vessels, 154
xylem parenchyma, 154
Pimpinella anisum
biological source, 23
family, 23
microscopy
cremocarp, 23
embryo, 25
endocarp, 24
endosperm, 24
epicarp, 24
mesocarp, 24
testa, 24
powder characters
endocarp, 25
endosperm, 26
epicarp, 25
fibrovascular tissue, 26
sclereids, 26
testa, 26
trichomes, 26
vittae, 25
Plantaginaceae, 109

Plantago ovata
biological source, 109
family, 109
microscopy
embryo, 110
endosperm, 110
epidermis, 109
pigment layer, 109
powder characters
embryo, 111
endosperm, 111
epidermis, 110
starch granules, 111
Podophyllum. *See Podophyllum emodi*
Podophyllum emodi
biological source, 143
family, 143
microscopy
cork, 143
cortex, 144
endodermis, 144
epiblema, 144
epidermis, 143
exodermis, 144
pith, 144
vascular bundles, 144
powder characters
calcium oxalate crystals, 146
cork, 145
epiblema and exodermis, 146
parenchyma, 146
sclereids, 145
starch granules, 146
vessels, 146
Polygonaceae, 161

Q
Quassia. *See Picrasma excelsa*

R
Radial longitudinal section (RLS)
quassia, 148
sandalwood, 165
Ranunculaceae, 16
Rauwolfia. *See Rauwolfia serpentina*
Rauwolfia serpentina
biological source, 155
family, 155
microscopy
medullary rays, 156
periderm, 156
secondary phloem, 156
secondary xylem, 156
powder characters
calcium oxalate crystals, 159
cork, 158
parenchyma, 158, 159
pericyclic fragments, 159
starch granules, 158
vessels and tracheids, 159
xylem fibres, 159
Rheum officinale
biological source, 161
family, 161
microscopy
cambium, 161–162
cork, 161
cortex, 161
medullary rays, 162
pith, 162
secondary phloem, 161
secondary xylem, 162
powder characters
calcium oxalate, 163
cork cells, 163
parenchyma, 163
starch granules, 163
vessels, 163
Rhubarb. *See Rheum officinale*
Rubiaceae, 63, 104

S
Sandalwood. *See Santalum album*
Santalaceae, 165
Santalum album
biological source, 165
family, 165
microscopy
radial longitudinal section, 165
tangential longitudinal section, 165
transverse section, 165–166
powder characters
fibres, 165
medullary rays, 166
vessels, 165
xylem fibres, 165
Saraca indica
biological source, 33
family, 33
microscopy
medullary rays, 34
parenchyma, 34
pericycle, 34
pericyclic fibres, 34
periderm, 33
Saraca indica (*cont.*)
phloem fibres, 33

phloem parenchyma, 34
sclereids, 34
powder characters
calcium oxalate crystals, 36
cork cells, 36
phloem fibres, 36
sclereids, 36
starch grains, 36
Scrophulariaceae, 81
Sectioning methods
freehand sectioning methods, 6–7
materials, 5–6
Senna. *See Cassia angustifolia*
Simaroubaceae, 148
Solanaceae, 38, 48, 178
Squill. *See Urginea indica*
Stramonium. *See Datura stramonium*
Strychnos nux-vomica
biological source, 135
family, 135
microscopy
aleurone grains, 136
endosperm, 136
oil globules, 136
plasmodesmata, 136
testa, 135–136
powder characters
endospermic cells, 138
testa, 138
trichomes, 138
Sudan dyes
procedure, 12, 13
results, 13
staining solution, 12

T
Tangential longitudinal section (TLS)
quassia, 148
sandalwood, 165
Terminalia arjuna
biological source, 28
family, 28
microscopy
cork, 28
cortex, 29
medullary rays, 29
secondary phloem, 29
powder characters
calcium oxalate crystals, 31
cork, 31
fibres, 31
medullary rays, 31
parenchymatous cells, 29, 31
pigment matter, 31
starch granules, 31
Toluidine Blue O (TBO)
results, 10
staining procedures, 10
stain preparation, 10
Transverse section (TS)
quassia, 148–149
sandalwood, 165–166
Tulsi. *See Ocimum sanctum*
Turmeric. *See Curcuma longa*

U
Umbelliferae, 23, 58, 94
Urginea indica
biological source, 174
family, 174
microscopy
lower epidermis, 175
mesophyll, 174–175
upper epidermis, 175
powder characters
calcium oxalate, 176
epidermal cells, 176
mucilage cells, 176
parenchyma, 176
vessels, 176

V
Vasaka. *See Adhatoda vasica*
Vinca. *See Catharanthus roseus*

W
Withania somnifera
biological source, 38
family, 38
microscopy
cork, 38
cortex, 39
ground tissue, 39
medullary rays, 39
phloem, 39
primary xylem, 39
secondary xylem, 39
powder characters
calcium oxalate crystals, 41
cork cells, 41
fibres, 41
starch grains, 41
vessels, 41

Z
Zingiberaceae, 99, 186
Zingiber officinale
  biological source, 99
  family, 99
  microscopy
    cork, 99
    cortex, 100
    endodermis, 100
    ground tissue, 100
  powder characters
    fibres, 102
    oleoresin cells, 102
    parenchyma, 102
    starch grains, 102
    vessels, 102

GPSR Compliance

The European Union's (EU) General Product Safety Regulation (GPSR) is a set of rules that requires consumer products to be safe and our obligations to ensure this.

If you have any concerns about our products, you can contact us on ProductSafety@springernature.com

In case Publisher is established outside the EU, the EU authorized representative is:

Springer Nature Customer Service Center GmbH
Europaplatz 3
69115 Heidelberg, Germany

Batch number: 10756376

Printed by Printforce, the Netherlands